A PRELIMINARY STUDY OF THE PALEOECOLOGY OF THE

AMISTAD RESERVOIR AREA

Assembled by

Dee Ann Story
Vaughn M. Bryant, Jr.

Final Report of Research Under the
Auspices of the
NATIONAL SCIENCE FOUNDATION
(GS-667)

June, 1966

FRONTSPIECE. View up Rio Grande taken from a point
about three miles upstream from Pecos
River. Mexico to the left, Texas to
the right.

TABLE OF CONTENTS

LIST OF FIGURES

LIST OF TABLES

iv

INTRODUCTION

Dee Ann Story

The primary purpose of this project has been to conduct a preliminary analysis of biological specimens recovered from archeological sites in the Amistad Reservoir area of southern Val Verde County, Texas, and northern Coahuila, Mexico (Figure 1). It is basically a feasibility study designed to evaluate the practicality and desirability of a more intensive, long-termed investigation into the paleoecology of that area.

The proposed Amistad Reservoir will be created by a dam presently being constructed by the International Boundary and Water Commission approximately 12 miles above the city of Del Rio, Texas, and one mile below the mouth of the Devils River (Figure 3). When completed, probably in 1968, this dam will begin backing water some 75 to 80 miles up the Rio Grande and will eventually extend 18 to 20 miles up the Pecos River and approximately 30 miles up the Devils River (Graham and Davis, 1958:1-2).

Characterized by a semi-arid climate and rolling uplands gouged by numerous steep-sided canyons (Figure 2), the Amistad region lies at the southern edge of Fenneman's (1931: 50-53) Great Plains physiographic province. Strictly speaking, the entire reservoir area falls within the Edwards Plateau subdivision. However, the uplands to the west of the Pecos River and just south of Rio Grande form a unit generally known as the Stockton Plateau (ibid.). The most outstanding geologic feature of the Stockton Plateau and the adjacent portion of The Edwards Plateau is the extensive outcrops of Cretaceous (Comanche series) limestone which are largely responsible for the rough, sharply eroded terrain and numerous caves and rockshelters dotting the canyon walls. A thin, almost non-existent, fluviatile mantle appears on the uplands while strips of alluvium occur along the Rio Grande and, to a much lesser extent, along the Pecos and Devils.

With an average annual rainfall of 18.13 inches, the region today is generally deficient in moisture (U.S. Dept. of Commerce, 1964). Except for three perennially flowing rivers--the Rio Grande, Pecos, and Devils--surface water is scarce throughout much of the year. Typically, the tributary canyons carry water only during the brief periods of rain. The average temperature varies from 51°F in January to 85°F in July. The maximum temperature recorded is 111°F and the minimum 11°F.

FIGURE 1. Map showing the location of the Amistad Reservoir and pertinent geographic features.

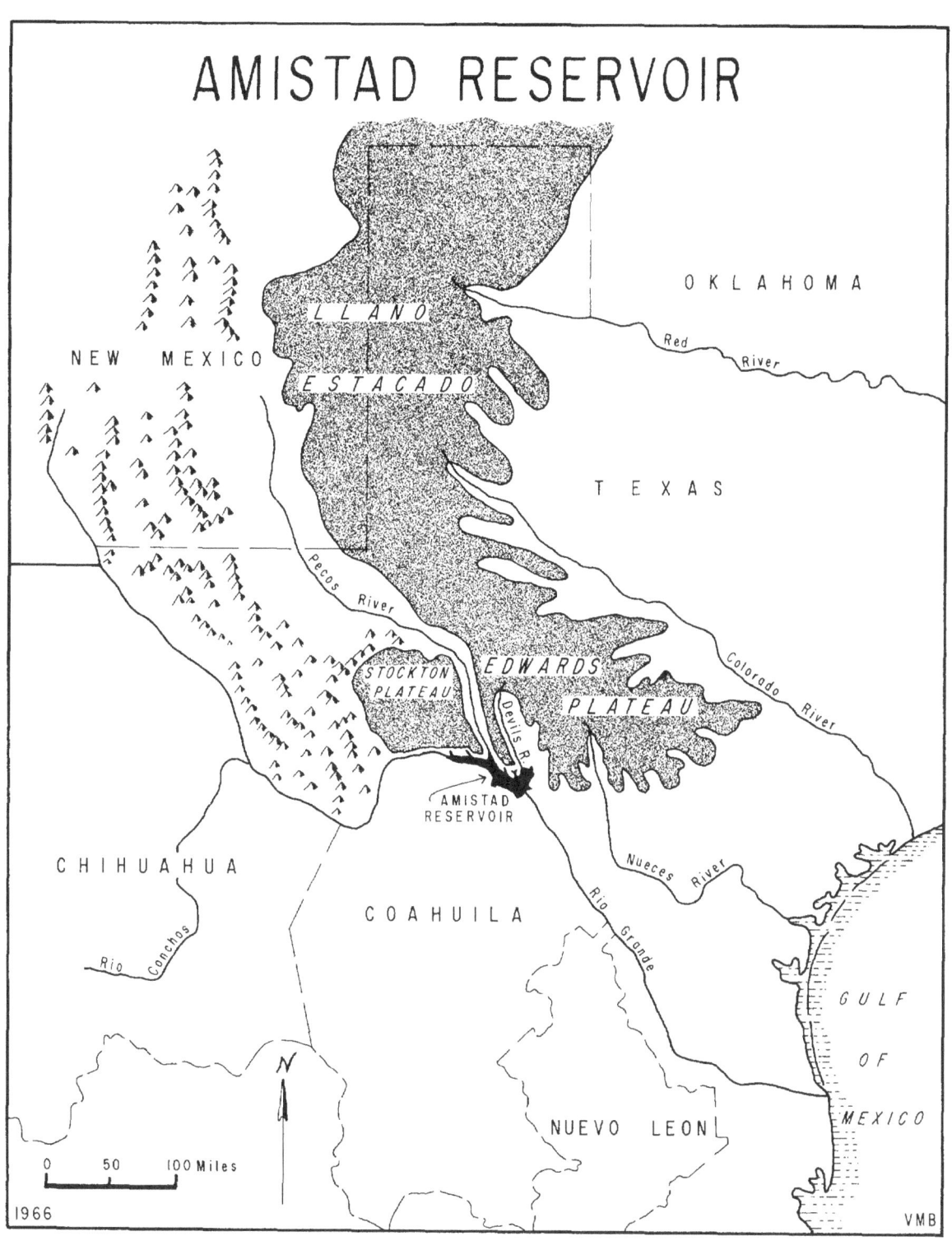

AMISTAD RESERVOIR

OKLAHOMA

NEW MEXICO

LLANO

ESTACADO

Red River

T E X A S

Pecos River

STOCKTON PLATEAU

EDWARDS

PLATEAU

Colorado River

Devils R.

AMISTAD
RESERVOIR

CHIHUAHUA

Nueces River

COAHUILA

Rio Grande

Rio Conchos

GULF

OF

MEXICO

N

NUEVO LEON

0   50   100 Miles

1966

VMB

The Amistad region lies at the critical juncture of three biotic provinces, the Tamaulipan, Chihuahuan, and Balconian (Blair, 1950). The fauna--as well as the flora--shows interdigitation of forms from these areas, particularly the Tamaulipan and Chihuahuan. Mammals common to the region include peccary, deer, coyote, jackrabbit, beaver, rock squirrel, skunk, fox, and ringtail. A variety of birds, reptiles, myriapods, and arachnids are also to be found.

In broad terms, the dominant vegetation consists of forms adapted to arid and semi-arid conditions (see Flyr's report for a more detailed statement of the flora). Most notable of the drought-resistant plants are yucca, sotol, lecheguilla, ocotillo, cat's claw, cenizo, nolina, cresote brush, and numerous cacti. In more favorable areas, such as along the permanent streams, there are motts of oak and a few pecans, walnut, hackberry, willow, and cottonwood trees. Mesquite and other legumes occur more widely but are nowhere to be found as abundantly as in some of the neighboring regions.

While to the casual visitor the Amistad Reservoir area appears as a hot, hostile environment, it is quite rich in evidence of prehistoric occupation. Caves and rockshelters afforded natural shelter from the elements, and the terraces along the permanent streams provided ideal camping spots with easy access to water. Additionally, the seemingly scant and scrubby vegetation contained many edible plant foods as well as usable parts for the manufacture of implements and household goods. Plants were clearly augmented by a variety of game animals.

In view of the terrain and the localized distribution of water, it is not surprising that the archeological sites are most concentrated in the sheltered portions of the canyons and along the terraces--the very areas to be most directly and immediately affected by the raising waters of the reservoir. The significance of this loss is much increased by the long record--at least 10,000 years--of prehistoric occupation preserved in the reservoir area and by the extraordinary condition of perishable items (cultural materials such as sandals, netting, basketry, throwing sticks, and the like, as well as numerous unworked plant remains) from the caves and rockshelters. This unusually lengthy and, from the archeologist's point of view, remarkably complete record offers a unique opportunity for paleoecology studies.

The task of retrieving archeological remains from the areas threatened by the Amistad Reservoir has fallen to the Texas Archeological Salvage Project (T.A.S.P.). This work, carried out in cooperation with Region Three of the National Park Service, is a part of the nationwide Interagency Archeological Salvage Program. Although the planned program of investigation is not yet completed, much valuable data have already been recovered. A meaningful chronology, supported by a number of radiocarbon dates, is being developed. Even more importantly, the nature, history, and dynamics of the aboriginal cultures are gradually being unraveled.

In addition to acquiring strictly archeological data, the salvage project archeologists have been collecting biological specimens and will continue to do so until the reservoir is completed, probably in 1969 or 1970. Although only preliminary, the studies reported here represent the first intensive analyses of these materials. Included are specimens from seven sites (Eagle Cave, Bonfire Shelter, Coontail Spin, Zopilote Cave, Devils Rockshelter, Devil's Mouth, and Fate Bell Shelter) excavated between 1961 and 1965.

The varied nature of the biological collections has necessitated an interdisciplinary approach. Dr. Donald A. Larson of the Botany Department of The University of Texas has supervised the botanical studies; Dr. Gerald G. Raun of the Texas Memorial Museum, The University of Texas, has directed the vertebrate zoological studies; Dr. E. P. Cheatum of the Department of Biology, Southern Methodist University, has been responsible for the analysis of the invertebrate fauna. I have, with the able assistance of David S. Dibble and Richard E. Ross, provided the archeological background. Others who have actively participated in this study are:

Akersten, William A.--graduate student, Univ. of Texas; assisted in the inventory of modern fauna.

Benfer, Alice N.--graduate student, Univ. of Texas; helped organize for study biological specimens from archeological sites.

Bryant, Vaughn M., Jr.--graduate student, Univ. of Texas; conducted analyses of pollen of Devil's Mouth and Devils Rockshelter sites; laboratory preparation of fossil pollen; assisted in preparation of pollen reference slides; drafted maps and pollen diagrams for final report.

FIGURE 2. The Amistad Reservoir area. a, View looking
down Pecos River toward Rio Grande. b, View
looking down Miles Canyon toward the Rio
Grande. c, View across uplands. d, View
looking down Rio Grande.

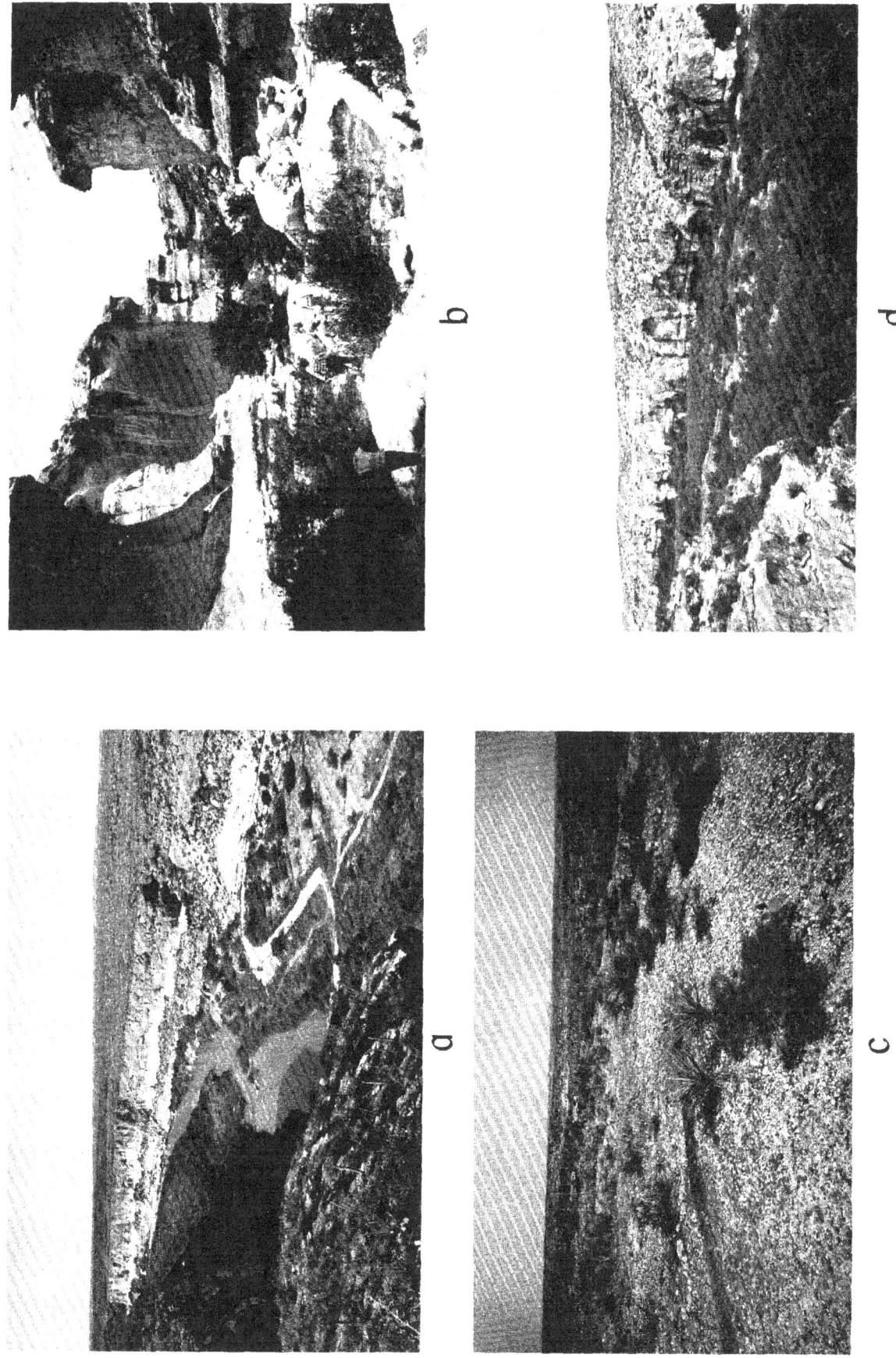

a

b

c

d

Burton, Sherry--typist; prepared final multilith masters.

Corbin, James E.--undergraduate student, Univ. of Texas; helped organize for study biological specimens from archeological sites.

Devine, Michael C.--undergraduate student, Univ. of Texas; assisted in the inventory of modern vertebrate fauna.

Eck, Lowell J.--research assistant; assisted in the taxonomic analysis of faunal remains from archeological sites and collection of modern fauna.

Flyr, Lowell David--graduate student, Univ. of Texas; collection and study of modern flora.

Greer, John W.--graduate student, Univ. of Texas; helped organize for study biological specimens from archeological sites.

Hevly, Richard H.--faculty associate, Humboldt State College (Arcata, Calif.); analysis of pollen from Bonfire Shelter.

*Jelks, Edward B.--faculty associate, Southern Methodist University; principal investigation of project, Dec. 23, 1964, to May 20, 1965.

*Jones, Melinda J.--secretary, Texas Archeological Salvage Project; secretarial duties.

*Kankrlik, John--undergraduate, Southern Methodist University; assisted in the collection of modern mollusk.

Irving, Robert S.--graduate student, Univ. of Texas; analysis of vegetal remains from archeological sites.

*Lazicki, Terrisa--clerk-typist, Texas Archeological Research Laboratory; assisted in typing drafts of manuscript.

Leonard, Cuyler H.--undergraduate student, Southern Methodist University; assisted in analysis of invertebrate fauna.

Litzler, Lee G.--undergraduate student, Univ. of Texas; assisted in the inventory of modern vertebrate fauna.

*Lorrain, Dessame--research associate, Southern Methodist University; analysis of vertebrate fauna from Bonfire Shelter.

---

*Salary not paid by grant

*McAndrews, John H.--faculty associate, Jamestown College
(Jamestown, North Dakota); analysis of pollen from
Eagle Cave; preparation of pollen reference slides;
preparation of pollen key.

Walters, Carol Ann--undergraduate, Univ. of Texas; assisted
in processing pollen.

*Wood, Barbara--secretary, Texas Archeological Salvage
Project; secretarial and bookkeeping services.

     In addition to persons mentioned above there are a
number of individuals who have assisted with various phases
of the project. Personnel from the Texas Archeo. Salvage
Project were cooperative and displayed much interest in the
study. Particularly to be thanked are Messers. David S.
Dibble, Richard E. Ross, Curtis D. Tunnell, Mark L. Parsons,
William Harrison, and Elton R. Prewitt, who provided useful
information concerning the archeology and assisted in the
collection of certain biological specimens. Dr. J. Richard
Ambler, Executive-Director of the T.A.S.P., generously made
available salvage project vehicles and other items of equipment. Technical assistance in the identification of vertebrate remains was provided by Dr. Ernest L. Lundelius, Jr.,
of the Department of Geology, The University of Texas.
Dr. Marshall C. Johnston of the Department of Botany, The
University of Texas, was helpful in the identification of
the modern flora. Comparative material for vertebrate
study was provided by The University of Texas Vertebrate
Paleontology Laboratory under the direction of Dr. John A.
Wilson, Professor of Geology. Comparative material for
the identification of the macrofossils and pollen to make
the reference slide collection were obtained from The
University of Texas Herbarium under the direction of Dr.
B. L. Turner, Professor of Botany.

_____

*Salary not paid by grant

FIGURE 3. Map locating sites included in study.

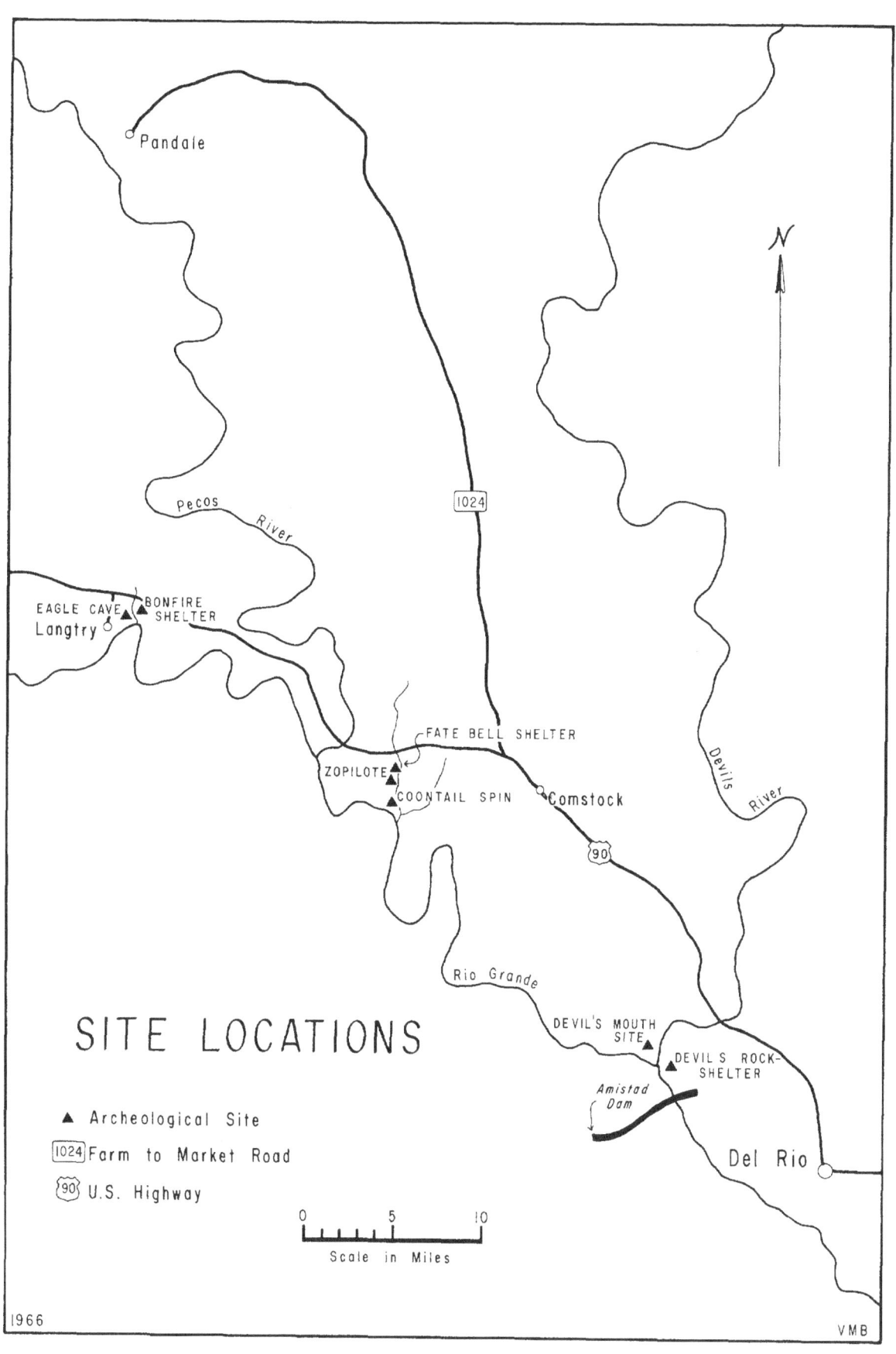

Pandale

1024

Pecos River

EAGLE CAVE ▲ BONFIRE
Langtry SHELTER

FATE BELL SHELTER

ZOPILOTE ▲
COONTAIL SPIN ▲   Comstock

Devils River

90

Rio Grande

DEVIL'S MOUTH
SITE ▲
▲ DEVIL'S ROCK-
SHELTER

Amistad
Dam

Del Rio

SITE LOCATIONS

▲ Archeological Site
1024 Farm to Market Road
90 U.S. Highway

0   5   10
Scale in Miles

1966   VMB

ARCHEOLOGICAL BACKGROUND

Dee Ann Story

Previous Investigations

The abundance and unusual preservation of specimens from sites in the Amistad area have long attracted the attention of both amateur, "relic," collectors and professional archeologists. A good deal of the early institutional work in the area, however, is poorly documented and the uncontrolled digging by artifact-seekers has destroyed--and is continuing to destroy--much valuable data.

The first serious excavations appear to have been made in 1932 when The University of Texas dug a small portion of Fate Bell Shelter (41 VV 74*) in Seminole Canyon (Pearce and Jackson, 1933; Thomas, 1933) and Gila Pueblo undertook limited excavations at Eagle Cave (41 VV 167) in Mile Canyon (Taylor, 1949b). These activities were quickly followed by Witte Memorial Museum's work at several caves near the now-abandoned town of Shumla (Martin, n.d., 1933; Schuetz, 1956, 1961, 1963) and by Smithsonian Institution's tests at two caves in the lower Pecos area (Setzler, 1934). Three years later, in 1936, The University of Texas dug at Horseshoe Cave (also known as Mrs. Martin Kelly Ranch Site; 41 VV 171) near Comstock (Butler, 1948) and Witte Memorial Museum carried out limited investigations at Eagle Cave. In 1937 the West Texas Museum excavated a portion of Murrah Cave (41 VV 351) some 25 to 30 miles up the Pecos River (Holden, 1937). In 1948 and 1949, Herbert C. Taylor, then a student at The University of Texas, did some surveying and testing in the area about the mouth of the Pecos River (Taylor, 1948, 1949a, 1949b).

Many of the early workers in the Amistad area noted the presence of numerous pictographs in the protected recesses of the shelters, but A. T. Jackson (1938) and Forrest Kirkland (1937, 1938, 1939) were the only ones to record them in

---

*The site designation for this and the other Amistad localities follows the system currently in use by the Texas Archeological Research Laboratory, The University of Texas. In this system 41 stands for Texas, VV for Val Verde County, and the number, such as 74, for a specific site within a county.

careful detail. Indeed, of all the early archeological research in Val Verde County, the work of these two men has been the most enduring and useful to subsequent studies.

Work of the Texas Archeological Salvage Project

It was not until the late 1950's, when the Amistad (then known as Diablo) Reservoir was proposed that a new era of scientific investigation began, at least on the Texas side of the reservoir.* Virtually all of this research, commencing with an intensive survey in 1958 (Graham and Davis, 1958) and continuing with large-scale excavations carried out at Arenosa Shelter (41 VV 99) this past fall (Dibble, personal communication), has been conducted by the Texas Archeological Salvage Project. To date, over 300 sites have been located and 24 have been either tested or rather extensively excavated. Of these, only seven are included in the present study, the selection being determined largely by the extent of the archeological investigations and the quantity as well as quality of biological specimens recovered. These sites are Fate Bell Shelter (41 VV 74), Coontail Spin (41 VV 82), Eagle Cave (41 VV 167), Devil's Mouth (41 VV 188), Zopilote Cave (41 VV 216), Bonfire Shelter (41 VV 218), and Devils Rockshelter (41 VV 264). Though seemingly few in number, they present a reasonably good picture of the prehistory of the region, at least as it is now understood.

The Archeological Findings

Several culture complexes have been defined for the area (Sayles, 1935; Kelley, et al., 1940) but without exception these have been based upon poor data and have failed to be useful concepts for the current phase of research. While it would be an oversimplification to view the prehistoric cultures as homogeneous and unchanging, there does appear to have existed in the area a persistent cultural pattern dominated by a hunting and gathering mode of subsistence. As an archeological cultural type it can be characterized as an Archaic tradition which shows, at least

---

*Unfortunately, very little archeological work has been done on the Mexican portion of the reservoir, there being only brief accounts published by Herbert C. Taylor (1948) and Walter W. Taylor (1958).

superficially, many similarities with the widespread Desert
Culture (Jennings and Norbeck, 1955; Jennings, 1956) of
western United States. Like the Desert Culture, it appears
basically to represent an intimate and delicate adaptation--
at a rather simple technological level--to an arid or semi-
arid habitat. From the archeological point of view, documentation and further definition of this apparent adaptation
is, of course, one of the primary aims of an intensive paleoecological study.

Within the Amistad Reservoir area, accumulations of
occupational debris occur in alluvial terraces, in rockshelters and caves, and on the uplands, generally along the
canyon rims and at upper reaches of the tributary canyons.
Most sites contain evidence of intermittent occupation, presumably by small, transient social groups organized perhaps
along kinship lines. Artifacts typical of the open (terrace
and upland) campsites, such as the Devil's Mouth Site, are
utilitarian implements fashioned by chipping or grinding
stone. Tools and ornments made from bone and shell also
occur but are less frequent. In the protected recesses of
many of the shelters and caves, the artifact inventory is
much enriched by items manufactured from perishable materials such as wood, fiber, and hide. Flat, stream-worn
pebbles covered with painted designs--evidently highly
stylized human figures--are likewise often recovered from
sheltered deposits. Highly conventionalized pictographs
of several styles adorn many of the shelter walls, estheti-
cally culminating in large, polychrome anthropomorphic
beings, possible shamans.

Of all the material objects, none has received as much
attention and systematic treatment as the projectile points
chipped from stone. Most have been classified into a variety
of named, formal categories, generally referred to as types
(Suhm and Jelks, 1962; Johnson, 1964; Parsons, 1965; Dibble, 1965; Nunley, et al., 1965; Ross, 1965). Since the
projectile points 1) have been dealt with in great detail,
2) occur in both open and protected sites, and 3) show considerable morphological change through time, they provide
a useful (though admittedly gross and preliminary) means of
correlating the cultural strata at the various sites. Stratigraphic data, coupled with radiocarbon dating and other
independent lines of evidence (such as paleontology, geo-
morphology, and palynology) make it possible to use projectile
points as general time markers, much like index fossils are
employed in geology.

TABLE 1. TENTATIVE CHRONOLOGY, AMISTAD RESERVOIR AREA

Time Periods	Estimated Dates	C-14 Dates	Diagnostic Point Types or Styles
VIII	A.D. 1600-		Metal arrow points
VII	A.D. 1600-1000	A.D. 1220-1020 (Tx-38, Centipede Cave)	Cliffton, Perdiz, Toyah, and other arrow points
VI	A.D. 1000-200 B.C.	A.D. 680-420 (Tx-151, Fiber Layer, Bonfire Shelter) A.D. 340-180 (Tx-194, Fiber Layer, Bonfire Shelter) A.D. 1040-760 (Tx-130, Fiber Layer, Bonfire Shelter)* --- 190-510 B.C. (Tx-76, Coontail Spin) A.D. 790-570 (Tx-81, Coontail Spin) A.D. 1540-1160 (Tx-77, Coontail Spin)*	Ensor, Frio, Paisano, and Figueroa
V	200-1000 B.C.	720-940 B.C. (Tx-106, Bonfire Shelter) 460-660 B.C. (Tx-131, Bonfire Shelter) --- 1880-2120 B.C. (Tx-79, Feature 1, Coontail Spin)* 2340-2620 B.C. (Tx-78, Feature 1, Coontail Spin)* --- 290-470 B.C. (Tx-192, Zone II, Fate Bell Shelter)*	Montell, Castroville, Shumla, Marshall, Marcos

Table 1 (Continued)

Time Periods	Estimated Dates	C-14 Dates	Diagnostic Point Types or Styles
IV	1000-2500 B.C.	1380-1540 B.C. (Tx-136, Stratum IIa, Eagle Cave) - - - - - - - - - - 1270-1490 B.C. (Tx-191, Zone III, Level I, Fate Bell Shelter) - - - - - - - - - - 3960-5200 B.C. (Tx-41, 36"-48" Centipede Cave)* A.D. 510-290 B.C. (Tx-39, 48"-59.5", Centipede Cave)* 2820-3440 B.C. (Tx-42, 23.5", Centipede Cave)*	Langtry, Almagre, Val Verde
III	2500-4000 B.C.	3470-3730 B.C. (Tx-117, Stratum IId, Eagle Cave) 2650-2930 B.C. (Tx-137, Stratum IId, Eagle Cave) 2520-2740 B.C. (Tx-196, Stratum IId, Eagle Cave) 2450-2690 B.C. (Tx-203, Stratum IId, Eagle Cave) - - - - - - - - - - 2140-2300 B.C. (Tx-193, Zone III, Level II, Fate Bell Shelter)	Nolan, Pandale
II	4000-7000 B.C.	4050-4270 B.C. (Tx-138, Stratum III, Eagle Cave)* 2290-2530 B.C. (Tx-195, Stratum III, Eagle Cave)* 4230-3990 B.C. (Tx-139, Stratum IV, Eagle Cave)*	Gower-like, Early Barbed, Bifurcated Stem, Uvalde (?)

Table 1 (continued)

Time Periods	Estimated Dates	C-14 Dates	Diagnostic Point Types or Styles
II (cont'd)		6660-6960 B.C. (Tx-107, Stratum V, Eagle Cave)	
		6580-6880 B.C. (Tx-108, Stratum V, Eagle Cave)	
		4570-4810 B.C. (Tx-109, Stratum V, Eagle Cave)	
		6470-6710 B.C. (Tx-140, Stratum V, Eagle Cave)	
		6660-6960 B.C. (Tx-141, Stratum V, Eagle Cave)	
		6550-6910 B.C. (Tx-197, Stratum V, Eagle Cave)	
		5070-5510 B.C. (Tx-152, Inter-mediate Horizon, Bonfire Shelter)*	
I	7000 B. C. -	7950-8750 B.C. (Tx-80, Area A, 12' below datum, Coontail Spin)	Plainview, Plainview golondrina, Plainview-like, Folsom, Angostura, Lerma
		8120-8440 B.C. (Tx-153, Bone Bed 2, Bonfire Shelter)	
		6820-7100 B.C. (Tx-128, Sq. #1, Baker Cave)*	
		6850-7310 B.C. (Tx-129, Sq. #1, Baker Cave)*	

*Association of sample questionable, or period assignment tentative.

Inasmuch as the prehistory of the Amistad region appears to be a long, continuous record of human occupation, the division of this chronicle into time periods is largely arbitrary and a matter of convenience. The scheme proposed here and presented in Table 1 is based chiefly upon stratigraphic information from sites dug by the T.A.S.P. Each of the eight periods here recognized is defined solely by certain groups of projectile points. Estimated ages are derived from radiocarbon dates run at the Radiocarbon Laboratory, The University of Texas (Tamers, et al., 1964; Pearson, et al., 1965). In Table 2 the various cultural strata distinguished at the seven sites included in the present study have been assigned to the appropriate time period or periods.

Individual Site Descriptions

In the brief site descriptions which follow, particular attention is given to the cultural and geologic zoning as these, along with the prevailing projectile point (largely dart point) styles, provide the basic context for the biological specimens.

Fate Bell Shelter

Measuring over 500 feet in length and 110 to 140 feet in maximum depth (Figures 4 and 5), Fate Bell is one of the largest, best known, and most impressive shelters in the Amistad Reservoir area. It lies on the west side of Seminole Canyon (Figure 3) several miles above the mouth of that canyon. The obvious richness of the extensive midden accumulations within the shelter have for many years enticed both professional and amateur archeologists so that much of the deposit today is badly disturbed. The University of Texas carried out fairly extensive excavations in 1932 and, although two reports have been prepared (Pearce and Jackson, 1933; Thomas, 1933), the results of this work have yet to be analyzed in adequate detail. Unfortunately, provenience data necessary for proper restudy of the wealth of materials collected by the University have been lost.

Between 1932 and 1963, when the Texas Archeological Salvage Project dug three small test pits (Parsons, 1965), Fate Bell has been unbelievably ravaged by irresponsible collectors. Indeed, one of the main objectives of the 1963 investigation was to determine whether or not sufficient undisturbed areas remained to justify more extensive

FIGURE 4.   View looking southwest down Seminole
            Canyon toward Fate Bell Shelter.

FIGURE 5.   Fate Bell Shelter:plan of excavation and
            Profile A.

FATE BELL SHELTER
41 VV 74

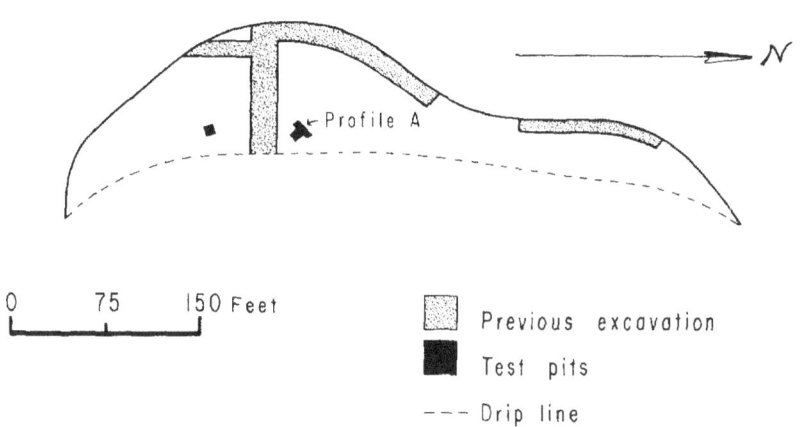

0   75   150 Feet

Previous excavation

Test pits

--- Drip line

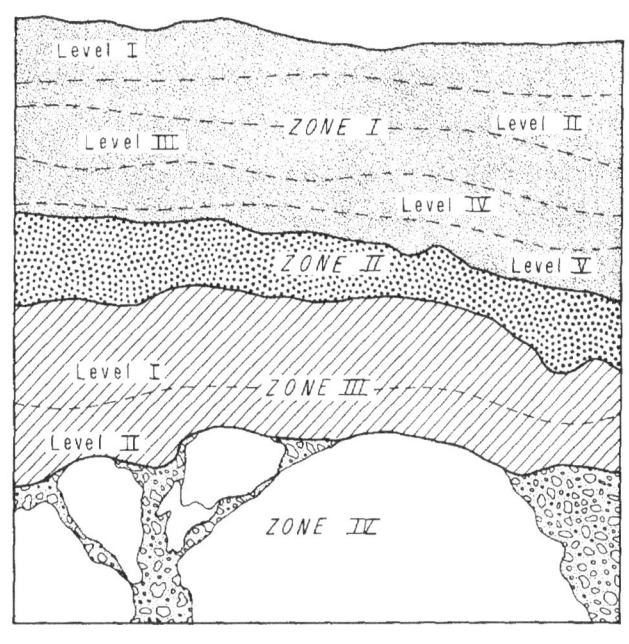

Level I

Level III ZONE I Level II

Level IV

ZONE II Level V

Level I ZONE III

Level II

ZONE IV

Profile A

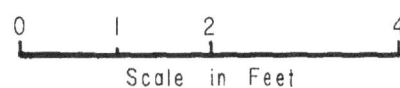

0   1   2   4

Scale in Feet

VMB   1966

Figure 6. Coontail Spin Site:plan of excavation and
profile showing stratigraphy in Area A.

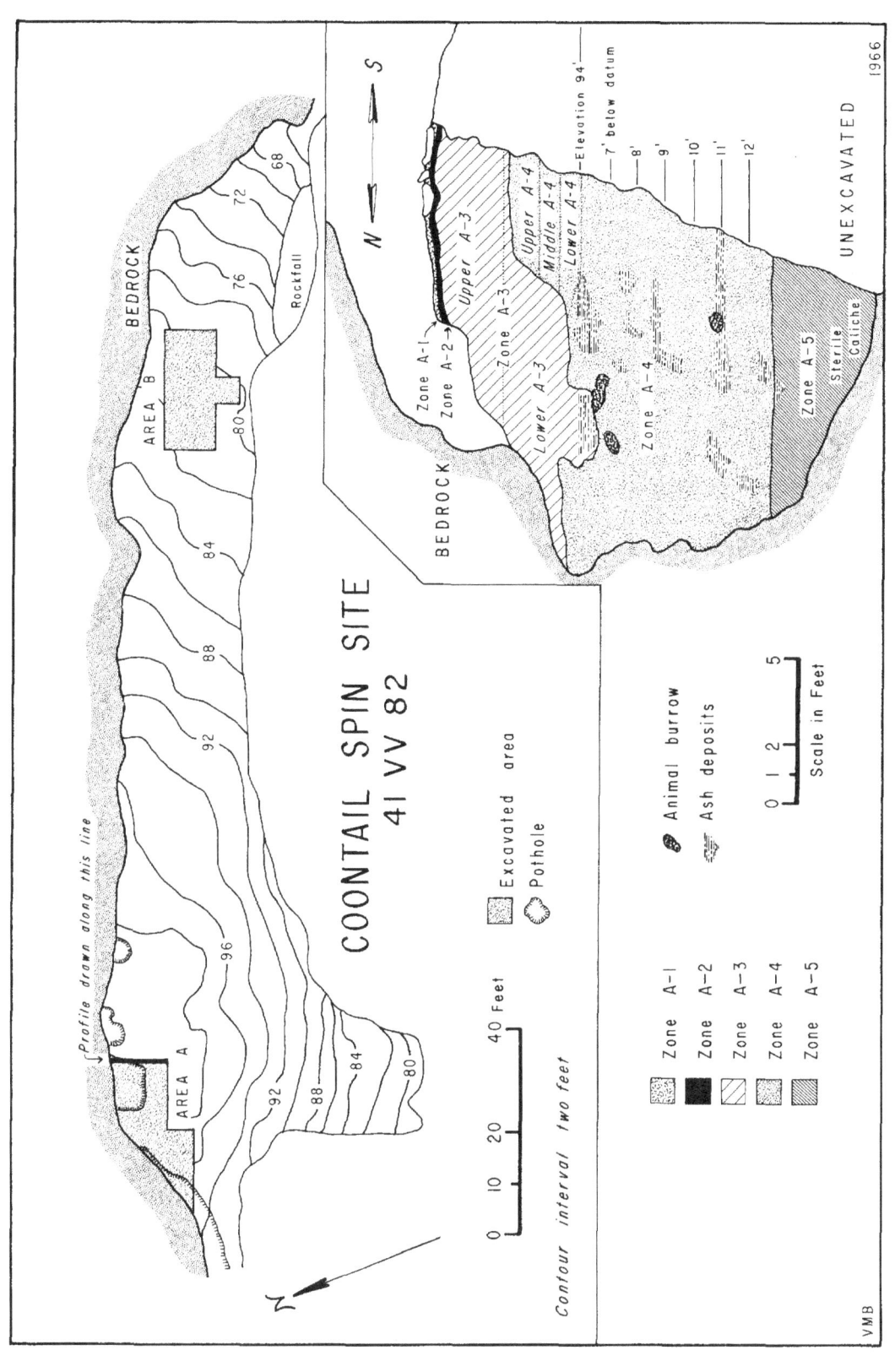

COONTAIL SPIN SITE
41 VV 82

Profile drawn along this line

AREA A

AREA B

BEDROCK

Rockfall

68
72
76
80
84
88
92
96

92
88
84
80

N

0   10   20   40 Feet

Contour interval two feet

Excavated area
Pothole

Zone A-1
Zone A-2
Zone A-3
Zone A-4
Zone A-5

Animal burrow
Ash deposits

0   1   2   5
Scale in Feet

N ← S

BEDROCK

Zone A-1
Zone A-2
Upper A-3
Upper A-3
Zone A-3
Lower A-3
Upper A-4
Middle A-4
Lower A-4
Zone A-4
Zone A-5
Sterile
Caliche

Elevation 94'
7' below datum
8'
9'
10'
11'
12'

UNEXCAVATED

VMB

1966

excavation by the T.A.S.P. in the future. From these tests
it was learned that some portions of the site are still more
or less intact and that a reasonably clear-cut stratigraphy
exists in the undisturbed parts of the deposit (Figure 5).
The zones and projectile point types found in the 1963 tests
include (Parsons, 1965):

Zone I, the uppermost stratum, was composed of
ash, burned stone and shell, and limestone dust. It
varied from dark to light gray in color and contained
five minor subunits or levels. Perishables were not
common, possibly having been destroyed by the extensive burning which had occurred in this layer. Dart
points, on the other hand, were relatively abundant
and consist mainly of Period VI styles, with types
Ensor and Frio being the most common. There are hints
of some mixture with earlier periods, but this appears
to be very slight.

Zone II, a black deposit comprised primarily of
charcoal, contained relatively few artifacts and no
distinct point forms. A C-14 date of 290-470 B.C.
(Tx-192; Pearson, et al., 1965) suggests that Zone II
could fall within Period V. In the absence of diagnostic cultural remains, and in view of the possibility
that this layer may represent a secondary deposit, this
period assignment must be regarded as quite tentative.

Zone III consisted of a brown-red soil which
contained a considerably quantity of vegetal material
as well as stones and snail shells. There were two
levels within Zone III with the uppermost, Level I,
having been a light brown soil apparently dating from
Period IV times. Both the point types (Langtry and
Val Verde) and radiocarbon date of 1270-1490 B.C. are
in agreement with this placement. Level II, distinguished mainly on the basis of its greater concentration of burned rock and decaying organic materials,
yielded one Pandale point and a radiocarbon date of
2140-2300 B.C. (Tx-193; Pearson, et al., 1965). It is
tentatively assigned to Period III.

Zone IV, the deepest stratum encountered during
the tests, was a yellow limestone dust which produced
a few artifacts along with numerous unburned stones
and roof spalls.

TABLE 2

ASSIGNMENT OF CULTURAL STRATA TO
CHRONOLOGIC PERIODS

Fate Bell Shelter (41 VV 74)

    Zone I - Period VI

    Zone II - tentatively, Period V

    Zone III, Level I - Period IV

    Zone III, Level II - tentatively, Period III

    Zone IV - tentatively, Period II

Coontail Spin, Area A (41 VV 82)

    Upper A-3 and Lower A-3 - mainly Period VI

    Transitional - mainly Period V

    Upper A-4 through Lower A-4 - mixed Periods III
                                          through V

    6 ft.-12 ft. below datum - mixed Periods I and II

Coontail Spin, Area B (41 VV 82)

    1 ft.-3 ft. below surface - mainly Period VI

    3 ft.-6 ft. below surface - mixed Periods IV and V

Eagle Cave (41 VV 167)

    Stratum I - Period VI with mixture from Periods VII
              and V through II

    Strata IIa and IIb - Period IV

15

TABLE 2 (Continued)

Strata IIc and IId - Period III

Strata III and IV - tentatively Period II, perhaps
extending into Period III

Stratum V - Period II

Devil's Mouth Site, Area A (41 VV 188)

Strata 1 through 5 - Periods VI and VII

Strata 6 through 9 - mainly Period V

Strata 10 through 12 - mainly Period IV

Strata 13 and 14 - mainly Period III

Strata 17 through 20 - Period II

Strata 21 through 24 - Period I, probably extending
into early Period II

Devil's Mouth Site, Area B (41 VV 188)

Surface-2.0 ft. - Period VII with slight mixture in
lower levels

2.0 ft.-5.5 ft. - mainly Period VI but also including
Periods VII and V

5.5 ft.-7.0 ft. - Period V

7.0 ft.-8.0 ft. - mixed Periods V, IV, and III

8.0 ft.-13.0 ft.- uncertain

Devil's Mouth Site, Area C (41 VV 188)

Upper Gravels - mainly Period I

TABLE 2 (Continued)

<u>Zopilote Cave</u> (41 VV 216)

> Artifact-bearing deposits undifferentiated but
> appear to represent Periods IV and V

<u>Bonfire Shelter</u> (41 VV 218)

> Fiber Layer - Period VI
>
> Bone Bed 3 - Period V
>
> Intermediate Horizon - tentatively, Period II
>
> Bone Bed 2 - Period I
>
> Bone Bed 1 - tentatively, Period I

<u>Devils Rockshelter</u> (41 VV 264)

> Zones VI-IX - uncertain
>
> Zones Ic through V - Period II
>
> Zone Ib through Id - uncertain

The one projectile point from this layer has not been
typed, but it appears to be similar to some of the
Period II points (the "Early Barbed") found by Richard
Ross at Eagle Cave (Ross, 1965). On this rather flimsy
evidence, Zone IV is hesitantly assigned to Period II.

Coontail Spin Site

This long (ca. 300 ft.) and rather narrow (maximum depth
of 40 ft.) shelter has been formed high in the north wall of
the Rio Grande canyon about three-quarters of a mile upstream
from the mouth of Painted Canyon (Figures 3 and 6). It was
first reported in the literature by Graham and Davis in 1958
and, fortunately, only a small section has been disturbed by
collectors. The surface of the fill within the shelter
slopes rather abruptly upwards from east to west, but there
are two relatively flat areas, one in the western end, the
other in the eastern part of the site. The salvage excavations at the site, carried out between September and November
of 1962, concentrated in these two areas (Nunley, et al., 1965:
3-14). As they cannot be stratigraphically linked, they have
been designated as separate areas; that in the western portion
is known as Area A, that in the eastern section as Area B.

Five different strata (Figure 6) were recognized in
Area A with only two (Zones A-3 and A-4) yielding man-made
objects. During excavation lower Zone A-3 and Upper A-4 were
mixed. Not actually a stratum, this mixed area is referred
to as Transitional. The three natural zones recorded at the
site in Area A include:

Zone A-1, a thin (two to five inches thick) super-
ficial layer of sheep dung and dust, yielded only a
small amount of cultural debris. Since the cultural
refuse appears to have been derived from materials
thrown out of potholes, Zone A-1 cannot be assigned to
a time period.

Zone A-2 was composed of a thin layer of white
limestone dust and immediately underlay Zone A-1. It
was devoid of artifacts and hence is not assigned to a
time period.

Zone A-3, the uppermost artifact-bearing stratum,
consisted mainly of a gray limestone dust from which a
considerable quantity of plant remains, perishable artifacts, and stone debris, and implements were recovered.
It was quite variable in thickness, reaching a maximum of about four feet in the center of Area A

and pinching out toward the rear of the site. To
provide an added measure of control, Zone A-3 was
arbitrarily subdivided into an upper and lower unit,
each comprising roughly half of the zone in any one
particular excavation. Dart points from both Upper
and Lower A-3 are mainly Period VI forms with the
Ensor type being the best represented. There is
some mixture with earlier periods, but, on the whole,
this is slight. The three radiocarbon dates (from
samples TX-76, 81, and 77) from A-3 are 190-510 B.C.,
A.D. 790-570, and A.D. 1540-1160 (Tamers, et al., 1964).
All but the last one appears reasonable for Period VI.

Zone A-4 consisted of a light-brown matrix dust
which was readily distinguished from Zone A-3. Artifacts, along with unworked items of detritus, were
found throughout the zone but were not generally as
numerous as in the preceding layer. Since A-4 was
a quite thick stratigraphic unit, it was subdivided
into a number of lesser divisions. That portion lying
above the 94 foot elevation (Figure 6) was divided
into three levels (Upper, Middle, and Lower). Below
this elevation, one foot intervals (i.e., 6-7 ft.,
7-8 ft., etc.) were used.

Dart points found in Upper A-4 through Lower
A-4 showed a good deal of variation and indicate this
portion of the deposit covers, at least in part, time
Period III through V. The two C-14 dates--1820-2120
B.C. (Tx-79; Tamers, et al., 1964) and 2340-2620 B.C.
(Tx-78; ibid.)--on two posts from a possible wind-
screen (Feature 1) assigned to the Upper A-4 (Period
V?) seem somewhat out of line and are of questionable
cultural association. The deeper section of A-4
(6 to 12 feet below the 100 foot datum elevation)
evidenced less mixture and can be assigned mainly to
Periods I and II. The deeper levels at Coontail Spin
contain the first definite evidence of actual occupation in a rockshelter during Period I. (At Bonfire
Shelter--see below--there are cultural remains
assignable to Period I, but these do not appear to
represent campsite debris.) One radiocarbon date of
7950-8750 B.C. (Tx-80; Tamers, et al., 1964) was obtained from the very lowest portion of Zone A-4 and
is of appropriate antiquity for Period I.

FIGURE 7. Views of Eagle Cave. a, Eagle Cave from across Mile Canyon. b, Interior of cave.

A

B

FIGURE 8. Plan map of Eagle Cave showing areas
           excavated and contour lines.

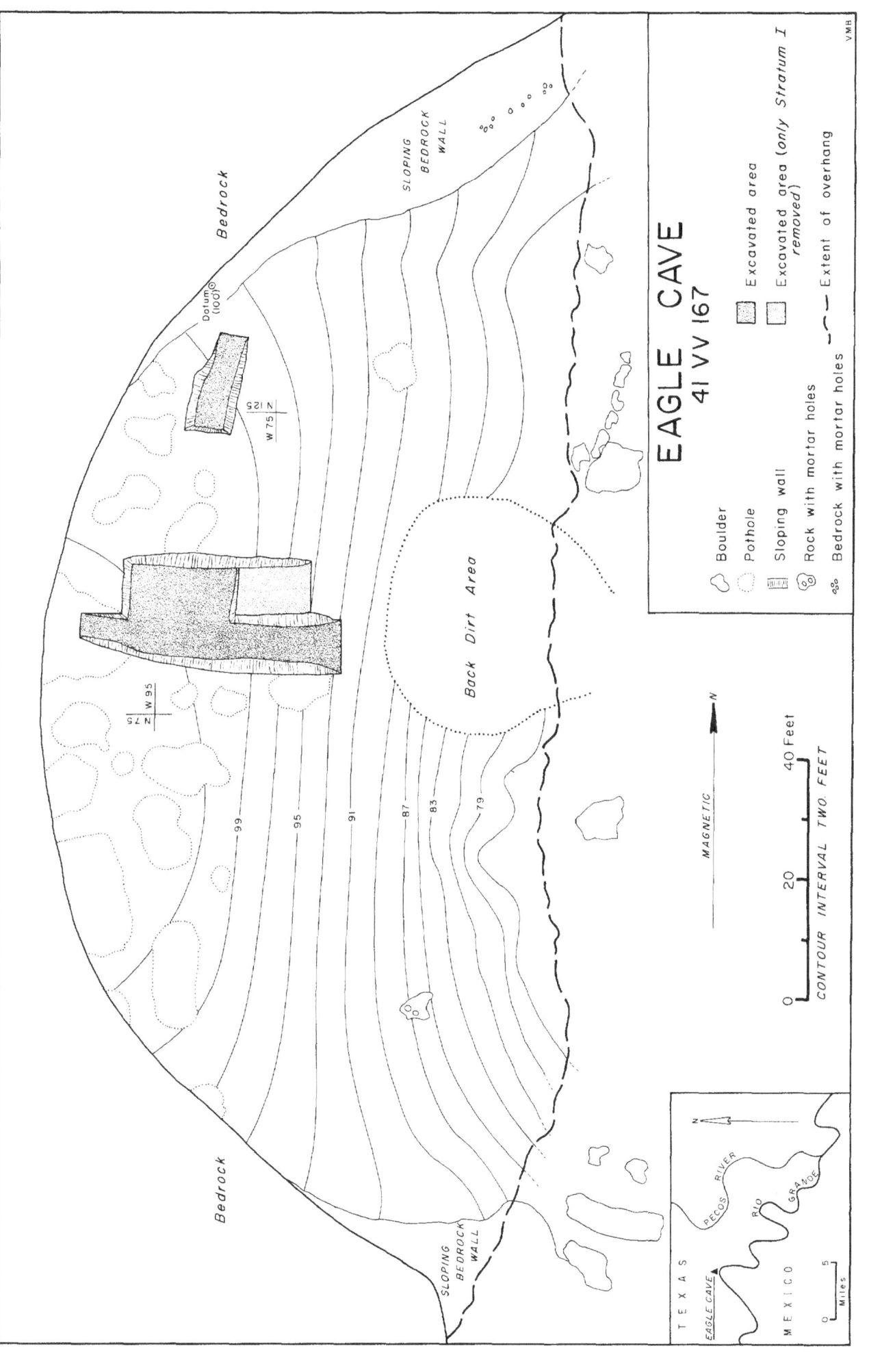

EAGLE CAVE
41 VV 167

Boulder
Pothole
Sloping wall
Rock with mortar holes
Bedrock with mortar holes

Excavated area
Excavated area (only Stratum I removed)
Extent of overhang

MAGNETIC N

0   20   40 Feet

CONTOUR INTERVAL TWO FEET

Bedrock

SLOPING BEDROCK WALL

Bedrock

Datum (100)

N 125
W 75

N 75
W 95

99
95
91
87
83
79

Back Dirt Area

SLOPING BEDROCK WALL

Bedrock

T E X A S

EAGLE CAVE

M E X I C O

PECOS RIVER

RIO GRANDE

N

0   5
Miles

VMH

Zone A-5 appeared to be culturally sterile as
only a few fragments of charcoal--possibly intrusive
from A-4--were recovered. It was composed of white
limestone dust varying from about one to three feet
in thickness. Limestone bedrock was encountered
immediately below Zone A-5.

All of the artifact-bearing zones at Coontail Spin
were extremely dry and the preservation of biological
specimens was extraordinarily good. The mixed Transitional layer, not discussed above, appears to date mainly
from Period V.

The stratigraphy in Area B of Coontail Spin was less
clear-cut and, as a result, the levels containing artifacts have been arbitrarily divided into two units. The
uppermost, one to three feet below the surface, contained
mainly Period VI points, while the lower member, three to
six feet below the surface, yielded specimens from both
Periods IV and V. In general, the upper portion of Area B
produced more artifacts than the lower (3 to 6 ft.) level.
Underlying the deepest cultural level there was stratum of
white limestone dust. It was lithologically similar to
A-5 and, like A-5, rested directly on bedrock.

Eagle Cave

Eagle Cave is a large, amphitheater-like opening on
the western side of Mile Canyon (Figures 7 and 8). It lies
approximately one-quarter of a mile from the Rio Grande,
but is rather difficult of access. Contained within the
shelter is an impressive accumulation of refuse which,
like that at Fate Bell, has been much disturbed. In addition to the 1963-64 T.A.S.P. investigations (Ross, 1965),
the site was dug in 1932 by E. B. Sayles and J. Charles
Kelley for Gila Pueblo (Taylor, 1949b) and in 1936 by J.
Walder Davenport for Witte Memorial Museum (Davenport, 1938).

Salvage project excavations at Eagle Cave (Ross, 1965)
revealed five artifact-bearing strata (Figure 9). Below
these units and extending to the limestone bedrock were
found thick (a total of at least nine feet) sterile deposits of spalls and gravel, which cannot be linked confi-
dently with a time period. The cultural layers recognized
at Eagle Cave included (from top to bottom):

Stratum I was a light soil containing burned
rock, animal dung, straw, and backdirt from older excavations. A much mixed, and partially artificial

zone, it contained numerous and varied artifacts
suggesting chiefly Period VI. However, points from
other periods, from VII and V to II, also occurred
and this assignment is of rather limited significance.

Stratum II was relatively undisturbed but quite
complex in composition, consisting of series of layers
which at times appeared distinct, at other times
faint. The upper portion, Stratum IIa, varied from
about a half to a foot and a half in thickness, was
grayish-brown in color, and contained large burned
rocks and fiber. Dart points most common to IIa were
those of Period IV--Langtry, Almagre, and Val Verde.
Tx-136, the one radiocarbon sample run from Stratum
IIa, yielded a date of 1380-1540 B.C. (Pearson,
et al., 1965).

Stratum IIb is a light gray layer immediately
beneath IIa. It was generally thin, rarely more than
half a foot in thickness, and was composed of small
rocks and limestone dust. Artifacts were less numerous than in the above zone but are also assignable
to Period IV.

Stratum IIc was a generally thin brown zone
containing a good deal of fiber and charcoal. Beneath
this was a complicated layer, IId, which showed a
great deal of variation in both color and composition.
Both strata contained dart point types (Nolan and Pandale) diagnostic of Period III. From IId four radiocarbon samples have been dated: TX-117, 3470-3730 B.C.;
Tx-203, 2450-2690 B.C.: Tx-137, 2650-2930 B.C.; Tx-196,
2520-2740 B.C. (Pearson, et al., 1965).

Stratum III was a relatively thick layer of fine
limestone particles. Charcoal was abundant in this
zone and two radiocarbon dates (Tx-138, 4050-4270 B.C.,
and Tx-195, 2290-2530 B.C.) have been obtained (Pearson, et al., 1965). Period assignment of this--and
the stratum below--is difficult perhaps because the
rather small projectile collection fails to fall
clearly into a pattern. The "Bifurcated Base" (Ross,
1965:53) may be the most diagnostic form although
"Early Barbed, Variety I" (ibid.:52), some Langtry,
and several nondescript points also occur. Suffice to
say that for the present Stratum III may represent late
Period II times, extending into Period III or IV.

FIGURE 9. East-west profile of Eagle Cave along North 90 line.

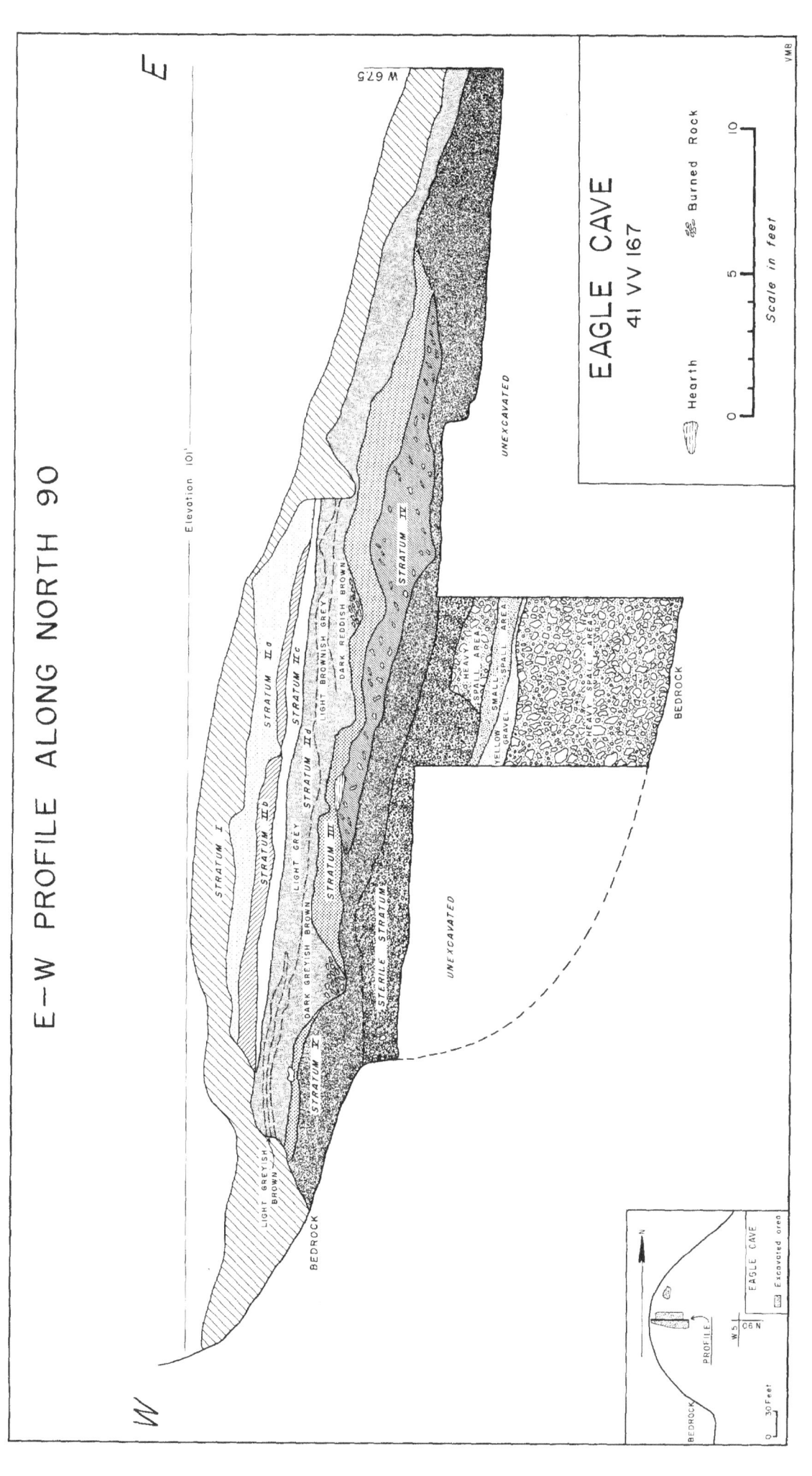

FIGURE 10. Devil's Mouth Site. a, View looking north
            across the Rio Grande. b, View looking
            northeast across the Rio Grande.

a

b

Stratum IV* was an easily recognized zone consisting of an accumulation of burned rock, ash, oxidized limestone, and a small amount of charcoal. It contained cultural remains similar in some respects to those found in Stratum III. Dart points, however, were less numerous and perhaps slightly less mixed. Stratum IV is very tentatively assigned to late Period II. The one radiocarbon sample (Tx-139) from this stratum yielded a date of 4230-3990 B.C. (Pearson, et al., 1965).

Stratum V, the deepest occupational layer found at Eagle Cave, was a light yellow-brown soil with small limestone spalls. Projectile points were not numerous but the dominant styles ("Early Barbed, Variety I and II") are clearly linked with Period II. The six radiocarbon dates from Stratum V range from 4570-4810 B.C. to 6660-6960 B.C.

Devil's Mouth Site

Lying at the confluence of the Rio Grande and Devils rivers, the Devil's Mouth Site consists of midden debris buried in an alluvial terrace remnant deposited against a high limestone bluff (Figures 10 and 11). This bluff raises some 130 feet above the Rio Grande and Devils, while the terrace surface is approximately fifty feet above these rivers. The relatively flat terrace is covered with a moderately heavy growth of brushes, weeds, and grasses. It extends some 150 feet southwest from the limestone cliff face towards the Rio Grande and for an undetermined distance along the Rio Grande. Occupational refuse can be observed on the terrace surface, along virtually all portions of the Devils River edge, and for about 1000 feet along the Rio Grande margin.

The site was first reported in 1958 by Graham and Davis, and excavations were carried out in December of 1959 (Johnson, 1961) and again in the fall and winter of 1961-62

---

*Due to a most regrettable error in the typing of Ross' report on Eagle Cave, a description of Stratum IV was omitted and Stratum V was mislabelled Stratum IV. The present characterizations of these two zones are based upon the descriptions in the original manuscript, not those in the published report.

(Johnson, 1964). All the excavations have concentrated in the eastern portion of the site, with those of the 1961-62 season being the most extensive. This latter work focused on three areas, A, B, and C (Figure 11). Area A was at the southeastern edge of the terrace, Area B five feet east of Area A, and Area C, farther to the east, at the base of the terrace.

Area A was excavated to a maximum depth of 36 feet below the surface and was found to contain a sequence of twenty-four (Strata 1-24) reasonably discrete zones (Figure 12). These were separated into two major units (Johnson, 1964:19) with Strata 1 through 13 making up the upper unit, and 14 through 24 the lower one. A marked erosional break separated the two and it is possible that there are some aeolian deposits in the upper unit. All portions of the lower unit appear to have been water-lain.

In Area B the excavations extended to a maximum of thirteen feet below the surface and the stratigraphy was less clear and possibly somewhat disturbed. Although they could not be sharply delineated, Strata 1 through 16 may also be represented in Area B. The zoning in Area C was clear enough but somewhat difficult to relate to that in the other parts of the site. The deposits here rested directly on the limestone bedrock, with the uppermost member consisting of a compact layer of tan sand with occasional lenses of silt. This zone possibly corresponds to Stratum 24 in Area A. Below the sand layer was found a zone of limestone and chert gravels, the "Upper Gravels," which varied from one to two and a half feet in thickness. Directly below the Upper Gravels were remnants of a second layer, the "Lower Gravels" which contained many igneous and metamorphic pebbles and cobbles. No gravels were encountered in Area A and it is possible that both the Upper and Lower Gravels pinched out toward the north.

Although cultural remains were recovered from most, but not all, of the strata at the Devil's Mouth Site, concentrated refuse was limited. In Area A, Strata 1, 3, 5, 7, 9, 11, and 13 were the main midden layers in the upper unit, while in the lower unit only Stratum 18 formed a recognizable occupation zone (Figure 12). Artifacts occurred throughout Area B but because of the vagueness of the stratigraphy they cannot be assigned to a specific stratum. In Area C, only the Upper Gravels produced cultural refuse.

Assessment of periods at the Devil's Mouth Site is as follows: In Area A, Strata 1-5 represent a mixture of

FIGURE 11. Plan view of Devil's Mouth Site showing
            contours and areas excavated.

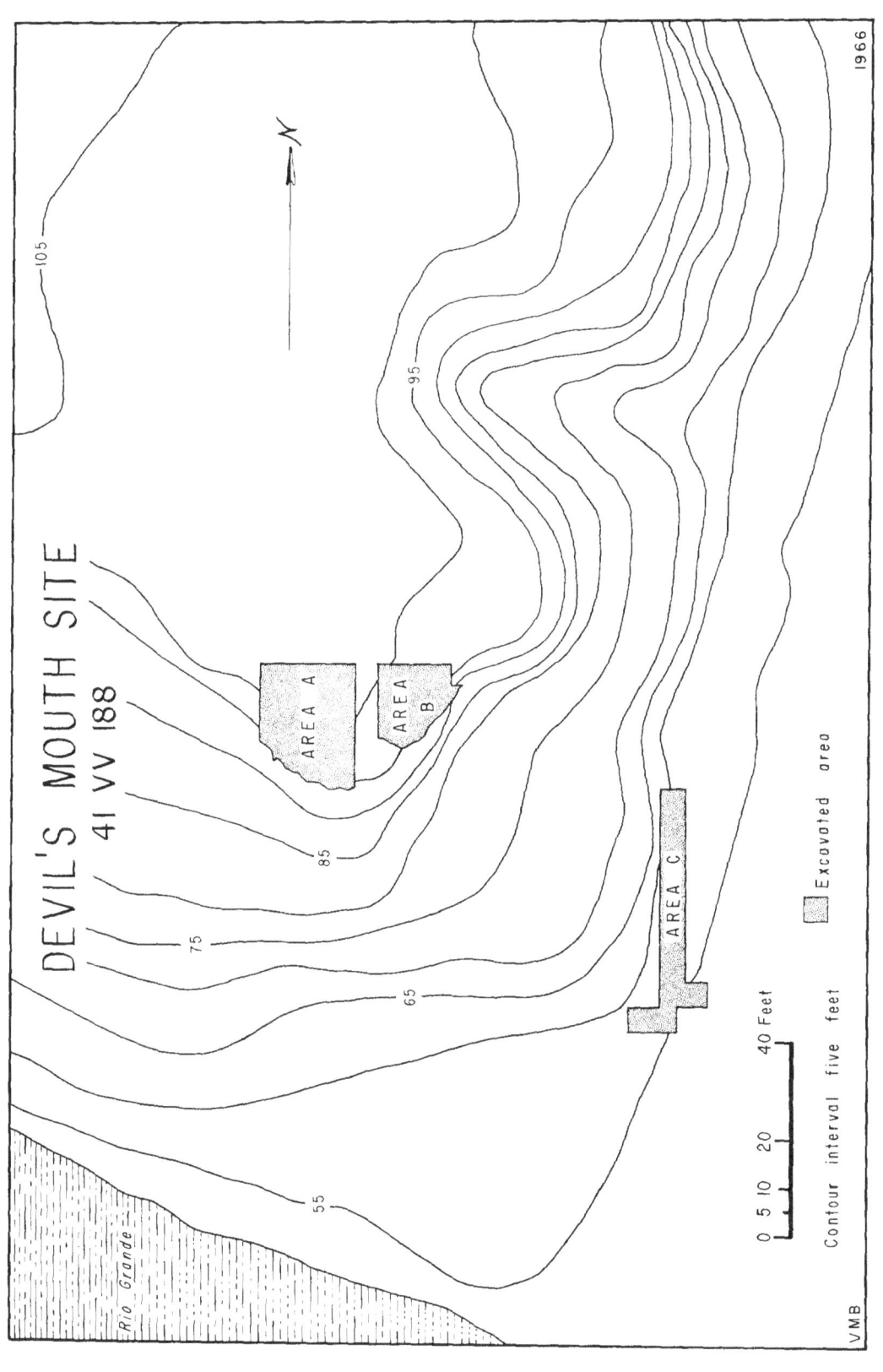

DEVIL'S MOUTH SITE
41 VV 188

Rio Grande

AREA A

AREA B

AREA C

Excavated area

0 5 10   20   40 Feet

Contour interval five feet

105

95

85

75

65

55

N

VMB

1966

FIGURE 12.   North-south profile of Devil's Mouth Site along W410.

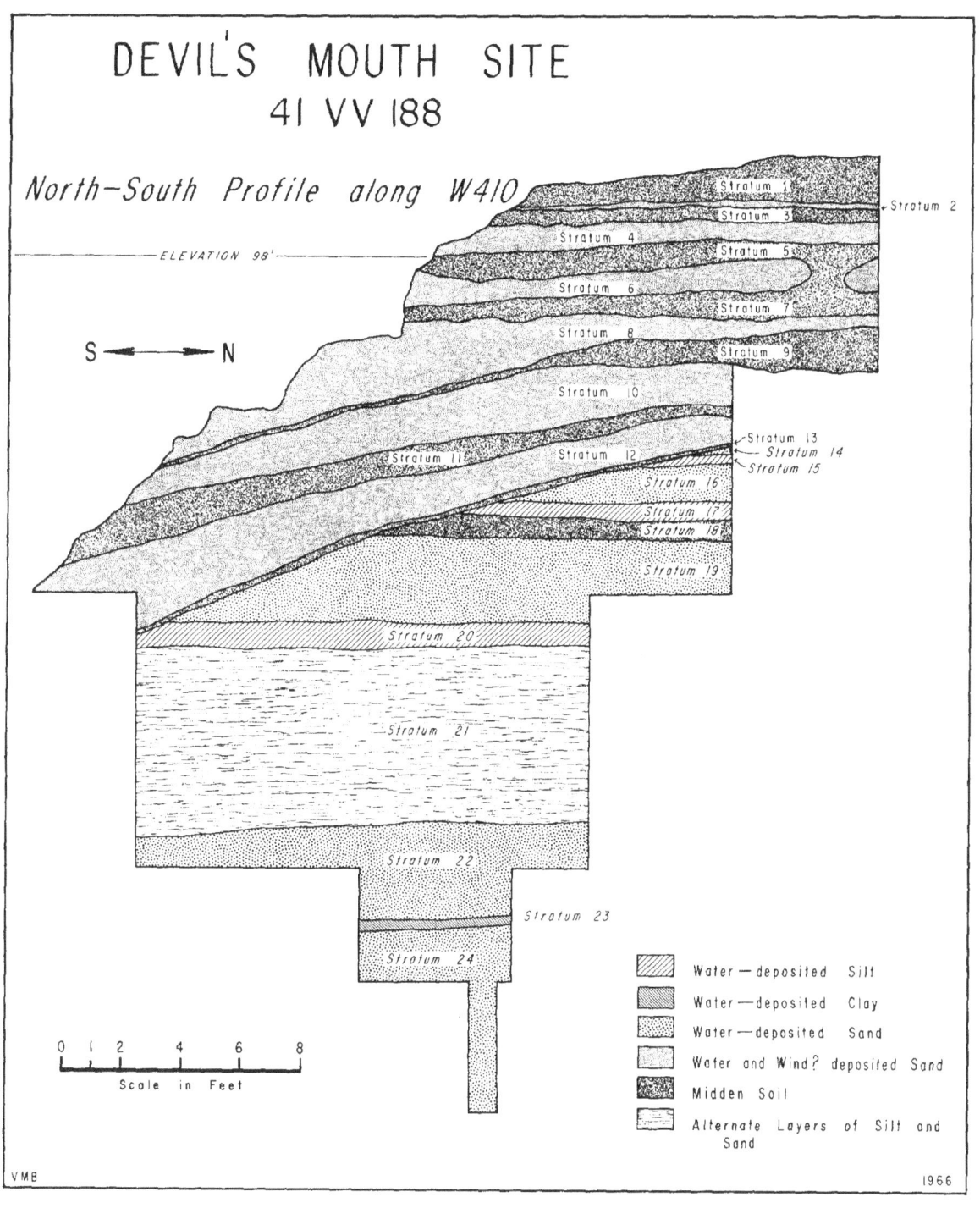

DEVIL'S MOUTH SITE
41 VV 188

North–South Profile along W410

ELEVATION 98'

S ◄———► N

Stratum 1
Stratum 2
Stratum 3
Stratum 4
Stratum 5
Stratum 6
Stratum 7
Stratum 8
Stratum 9
Stratum 10
Stratum 11
Stratum 12
Stratum 13
Stratum 14
Stratum 15
Stratum 16
Stratum 17
Stratum 18
Stratum 19
Stratum 20
Stratum 21
Stratum 22
Stratum 23
Stratum 24

0 1 2   4   6   8
Scale in Feet

Water—deposited Silt
Water—deposited Clay
Water—deposited Sand
Water and Wind? deposited Sand
Midden Soil
Alternate Layers of Silt and Sand

VMB   1966

Periods VI-VII; Strata 6-9 represent mainly Period V; Strata
10-12 represent mainly Period IV; Strata 13 and 14 represent
mainly Period III; Strata 17-20 represent Period II; and
Strata 21-24 possibly represent Period I extending perhaps
into Period II. In Area B, the surface to 2.0 ft. below the
surface represents Period VII with slight mixture in the
lower levels; 2.0 to 2.5 ft. represents Period VI with some
mixture with Periods VII and V; 5.5 to 7.0 ft. represents
Period V; 7.0 to 8.0 ft. represents mixed Periods V-III;
and 8.0 to 13.0 ft. cannot be assigned because of the paucity of dart points. In Area C, the most common points
(Plainview golondrina, Angostura, and, less certainly,
Lerma) are assignable to Period I.

Zopilote Cave

Zopilote Cave is a small rockshelter, approximately
60 feet long and 20 feet wide (Figure 13, b) discovered
during the 1962 T.A.S.P. field season. It lies in the wall
of a large arroyo that enters Seminole Canyon from the west
about 2 miles north of the Rio Grande. A large rockfall has
blocked the view of the shelter from the arroyo floor and
it is not surprising that Zopilote was overlooked by earlier archeological surveys and by collectors.

The 1962 salvage project excavations at the site were
limited so that the full nature of the deposits are not
known. The deepest pit extended to about six feet below
the surface, being halted after encountering some two feet
of relatively unproductive fill. On the whole, the deposits
were dry and contained numerous fragments of burned rock,
including one quite notable accumulation on the surface
(Figure 13, a).

Three zones, A, B, and C, were recognized with the
uppermost (Zone A) containing most of the occupational
detritus. This layer consisted of a quite dry, ashy soil
mixed with numerous items of debris. The two lower levels,
B and C, were composed of limestone dust, with C being
rather damp.

Cultural debris from Zopilote appears to be rather
mixed making it impossible to divide the deposit into
specific time periods. Most of the identifiable points
nonetheless seem to be styles characteristic of either
Period IV or V.

Bonfire Shelter

Located in the east wall of Mile Canyon about a quarter of a mile above Eagle Cave, Bonfire Shelter is one of the most unusual sites yet to be investigated in the Amistad area. Most unexpectedly, the excavations at this site uncovered two massive bison bone beds, apparently the results of driving herds of these animals off the overhang above the shelter. Bonfire was not found during the initial survey of the reservoir (Graham and Davis, 1958), as much of it is effectively obscured from the canyon floor by a massive rockfall (Figure 14). In 1962, the Salvage Project made limited tests and the extraordinary nature of the site was partially revealed. Additional and quite extensive excavations were carried out in the fall and winter of 1963-64 (Dibble, 1965).

Bonfire is a long (320 feet), narrow (maximum of 60 ft.) shelter which curves at each end (Figures 15 and 16). It faces to the west and all but the ends are blocked by the collapsed roof debris. The age of the collapse is unknown, but it clearly predates the earliest evidence of human use of the shelter. The surface of the deposits within the shelter are irregular and, in general, slope downward from the mass of ceiling debris (Figure 16). At the southern end of Bonfire there was a small talus cone--an accumulation of material below a notch or cleft in the canyon rim (Figure 14). It is in this mounted portion of the deposit that the bones were most concentrated. The evidence uncovered in this portion of the site most clearly revealed that bison had been driven into the notch and tumbled onto the under-lying talus cone surface.

The stratigraphy at Bonfire was, on the whole, remarkably distinct (Figure 17). It consisted basically of three major natural zones interspersed with four, possibly five, layers of cultural association (Dibble, 1965). The natural zones at the site include (from top to bottom):

Zone 3, the uppermost stratigraphic unit, was a somewhat variable stratum composed primarily of silt and limestone spall. In the southern portion of the site (about the talus cone) it consisted of unsorted spall and silt while in the more central section of the shelter it was composed primarily of light brown silt.

25

FIGURE 13. Zopilote Cave. a, View inside shelter looking southeast. b, Site plan showing contours and areas excavated.

ZOPILOTE CAVE
41 VV 216

a

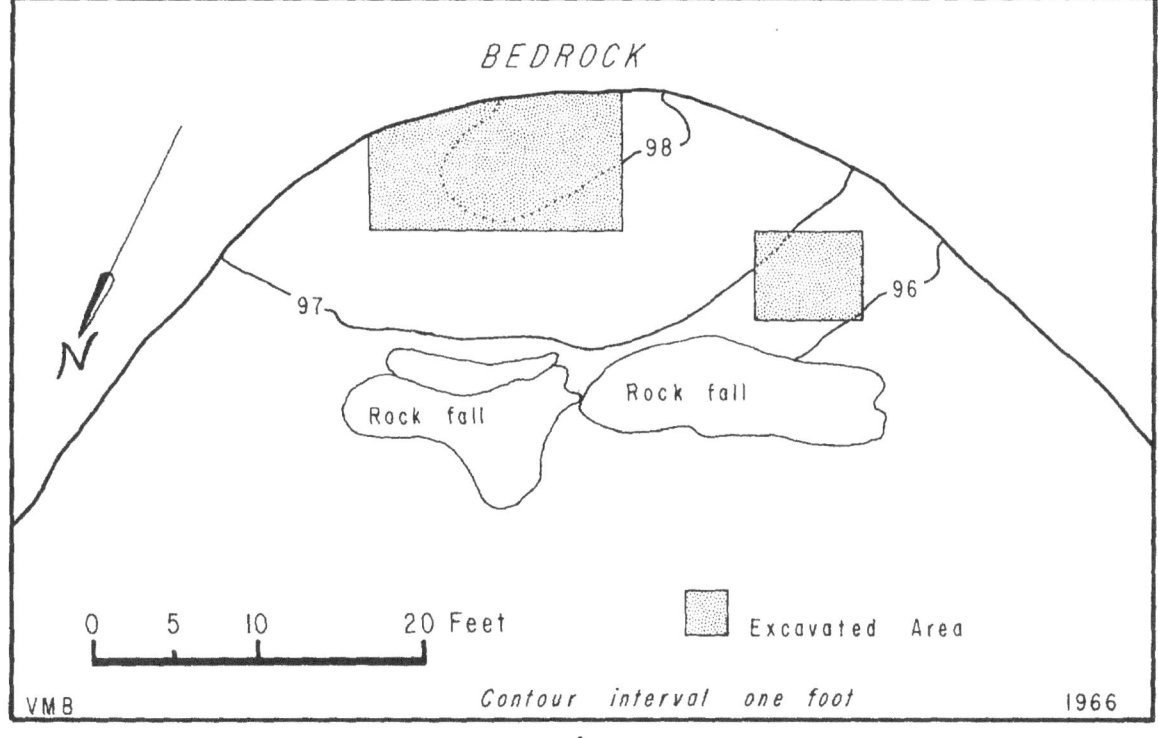

BEDROCK

98

97

96

Rock fall   Rock fall

0   5   10   20 Feet

Excavated Area

VMB   Contour interval one foot   1966

b

Zone 2 was a thick, rather heterogeneous deposit below Zone 3. The horizontal variation exhibited by this stratum was marked. As exposed on the southern flank of the talus cone it appeared as an undifferentiated tan to light gray accumulation of silt and limestone spall. Moving northward from the N50 line two distinct components (2a and 2b) could be discerned. The upper unit, Zone 2b, was a light brown layer composed largely of silt. Zone 2a was a tan to light brown layer of highly weathered spall and silt which rested directly on Zone 1.

Zone 1, the deepest encountered in the excavations, was a light gray deposit made up chiefly of weathered limestone spall. It was most clearly seen in the better sheltered areas of the site. At the southern, more open portion of the site, what was believed to be an equivalent layer showed considerably more weathering of the spalls and a generally higher silt content.

Within the natural deposits at Bonfire there were four definite layers of material associated with cultural remains. A fifth layer, Bone Bed 1, is suspected to be the result of human activity (Dibble, 1965). These cultural strata include (from top to bottom):

The Fiber Layer, composed primarily of a mass of vegetal debris, was concentrated at the southern end of the site. It occurred within Zone 3 and that portion of Zone 3 above the Fiber Layer was designated 3b, that underlying it was called 3a. This cultural stratum represents the only significant use of the shelter as a dwelling or camping spot. Hence the artifacts show more variation than those from the other cultural strata at the site. Points were not numerous but include the Ensor type which is assigned to Period VI. The Fiber Layer yielded two radiocarbon dates, A.D. 680-430 (Tx-151; Pearson, et al., 1965) and A.D. 340-180 (Tx-194; ibid.). A third date, A.D. 1040-760 (Tx-130; ibid.) may also apply to the Fiber Layer but the association of the sample is not certain.

Bone Bed 3, the uppermost accumulation of bison bones (all Bison bison), was found sandwiched between Zones 2 and 3 (Figure 18). It extended horizontally from the outer flanks of the talus cone, where it was

thickest, to the center of the shelter where it
consisted of only a scattering of bones. Certainly
the single most impressive feature at the site, Bone
Bed 3, is estimated to contain the remains of at
least 800 individual bison (see report herein by
Dessamae Lorrain). No significant breaks were dis-
cernible within the bone mass and it remains uncertain
as to the number of individual bison jumps involved.
It does, however, seem quite apparent that Bone Bed 3
accumulated over a relatively short period of time.
In the area of its most concentrated occurrence (on
the crest and flanks of the talus cone), Bone Bed 3
showed signs of having been intensively burned; the
intensity of the burning diminished east and north
of the cone. Cause of the fire remains undetermined.

Found associated with the bison bones were lithic
artifacts consisting mainly of projectile points
(Montell and Castroville-like) assignable to Period V.
There is no evidence indicating use of the shelter as
a camping site and all of the artifacts are objects
which can easily be linked with killing and butchering
of animals. The two C-14 dates from Bone Bed 3
(Pearson, et al., 1965) are 720-940 B.C. (Tx-106) and
460-660 B.C. (Tx-131).

The Intermediate Horizon was found in Zone 2 and
2a, primarily on the flanks of the talus cone and in
the rear center of the shelter. It contained rather
diffuse signs of light occupation including two
hearths, a few nondescript lithic artifacts, and
scattered flecks of charcoal. There were no diagnostic points which permit clear periodization; however,
the relative stratigraphic position of the Intermediate
Horizon and a radiocarbon date of 5070-5510 B.C.
(Tx-152; Pearson, et al., 1965) suggest it might well
fall within Period II.

Bone Bed 2 was the second and deepest accumulation
of bison bones found at the site (Figure 19). Its relationship to the natural zones was quite complex and
it will suffice to note here that in the central part
of the site it appeared immediately below Zone 2a,
north of the talus cone it appeared below Zone 2b, and
in the cone area it appeared at the base of undifferentiated Zone 2. Within this last-mentioned area Bone
Bed 2 could be divided into three components, A, B,
and C. Though somewhat less extensive than Bone Bed 3,
it had roughly the same horizontal extent. An estimated

FIGURE 14. Aerial view of Bonfire Shelter. Sheltered area is largely hidden by debris from massive collapse of canyon rim. Restricted entryways at north (left) and south (right) ends are visible. Note cleft in canyon rim above south end of sheltered area. Rio Grande is visible in the background.

FIGURE 15. Bonfire Shelter. a, View of central and
            southern parts of shelter prior to excavation.
            b, View into central portion of shelter.

a

b

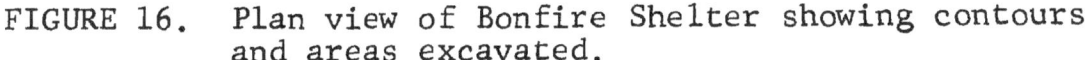

FIGURE 16. Plan view of Bonfire Shelter showing contours and areas excavated.

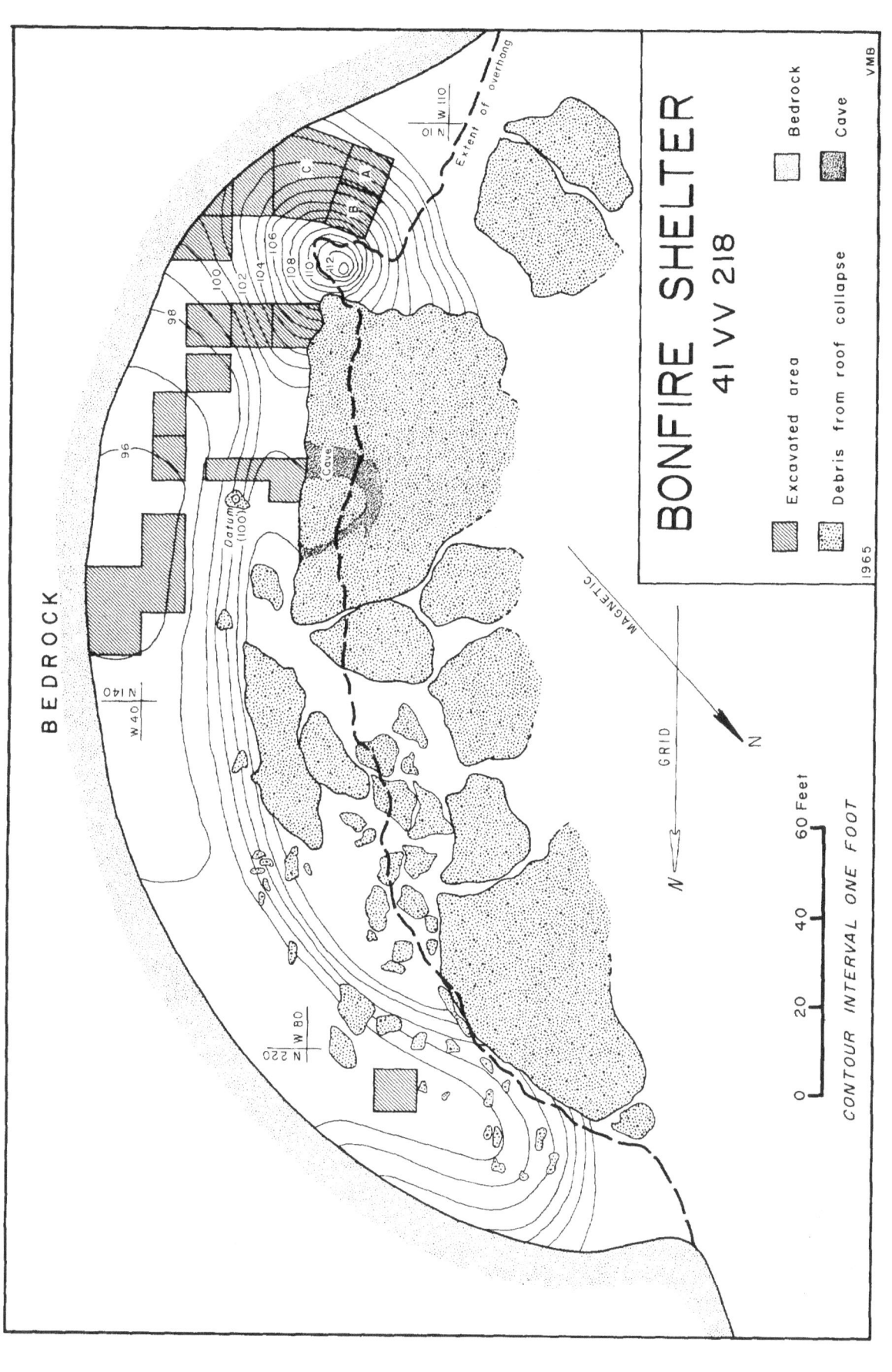

BEDROCK

BONFIRE SHELTER

41 VV 218

Excavated area

Debris from roof collapse

Bedrock

Cave

1965

VMB

N GRID

N MAGNETIC

0   20   40   60 Feet

CONTOUR INTERVAL ONE FOOT

Extent of overhang

N 10   W 110

Cave

Datum (100)

96

98

100

102

104

106

108

110

112

N 140   W 40

N 220   W 80

FIGURE 17.   Profile along north wall of Pits B, C, and
            Square N30/W60, Bonfire Shelter.

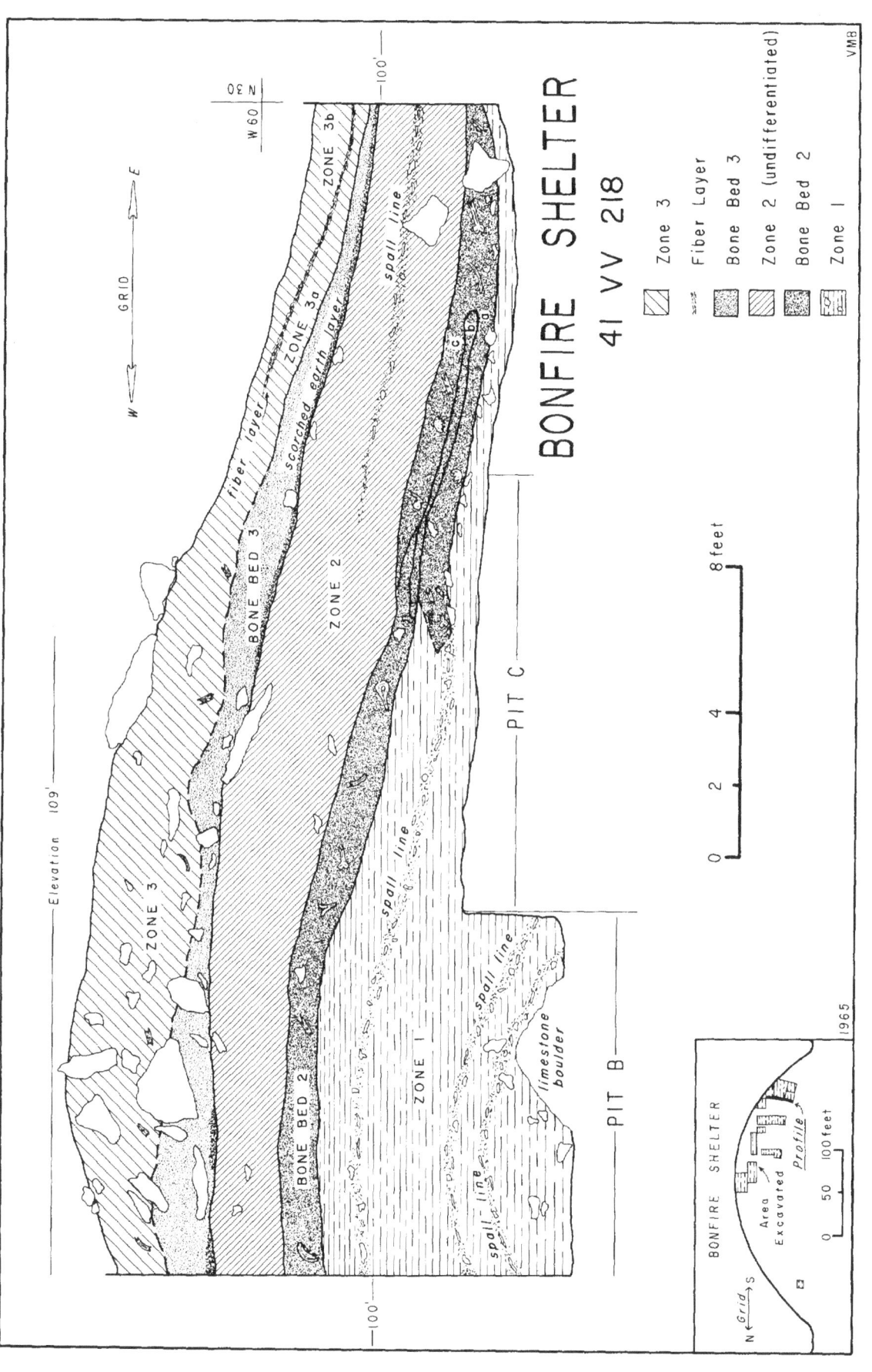

BONFIRE SHELTER

41 VV 218

Zone 3

Fiber Layer

Bone Bed 3

Zone 2 (undifferentiated)

Bone Bed 2

Zone 1

8 feet

0   2   4

Elevation 109'

ZONE 3

BONE BED 3

ZONE 3a

ZONE 3b

fiber layer

scorched earth layer

ZONE 2

spall line

BONE BED 2

ZONE 1

spall line

spall line

spall line

limestone boulder

PIT B

PIT C

100'

100'

N 30

W 60

W   GRID   E

BONFIRE SHELTER

N ← Grid → S

Area Excavated

Profile

0   50   100 feet

1965

VMB

FIGURE 18.   Views of Bone Bed 3, Bonfire Shelter. a, Photo
             of central portion of shelter showing maximum
             extent of excavations here. Visible is Bone
             Bed 3 (light-colored layer on far wall of trench
             in foreground) as it gradually lenses out toward
             figure in background. b, Skeleton of bison calf
             --the most complete articulation found during
             excavation of the bone beds.

a

b

FIGURE 19. Vertical view of portion of Bone Bed 2,
            Bonfire Shelter.

28

120 individual bison (all apparently of now-extinct form) are represented by Bone Bed 2, and at least three different jumps or drives were responsible for the accumulation. The faunal remains and matrix surrounding Bone Bed 2, like Bone Bed 3, showed signs of having been burned.

The artifact collection from this stratum is rather small, consisting of 21 lithic implements (projectile points, bifaces, scrapers, and utilized flakes) which were evidently used to kill and butcher the bison. All points are of Period I forms and can be classified as either the Plainview or Folsom type. The one Folsom point, significantly, was recovered from Component A, the earliest of the subdivisions recognized in Bone Bed 2. Within Bone Bed 2 there were no indications that Bonfire had been used as an occupational site. The one charcoal sample dating from this layer yielded an age of 8120-8440 B.C. (Tx-153; Pearson, et al., 1965). It is the oldest C-14 date yet obtained for cultural remains in the Amistad area.

Bone Bed 1 was found only in the central portion of the shelter, so that its full nature and origin are still open to question. It appeared as a relatively thin layer of large bones beneath Zone 2a and on top of Zone I. Included among the faunal remains were now-extinct forms of elephant, camel, horse, and bison. This assortment of bone, their disarticulated arrangement, and the presence of flecks of charcoal as well as large spalls (which may have been used to break the bones) all strongly suggest that Bone Bed 1 is a result of human activity. It is thus tentatively placed in Period I.

Devils Rockshelter

The Devils Rockshelter lies buried in alluvial deposits at the base of a slight overhang (Figure 20). It lies on the eastern side of the Rio Grande canyon just downstream from the mouth of the Devils River. As at the nearby Devil's Mouth Site, the terrace surface is about 50 feet above the waters of the Rio Grande. It is approximately 40 feet from the top of the bluff to the surface of the site, and the maximum extent of the overhang is only 10 feet.

When the T.A.S.P. conducted limited excavations in the spring of 1965, much of the terrace surface had already been damaged (Prewitt, 1966). It is estimated that approximately eight feet of overburden had been removed and used in the construction of the closeby, but non-abandoned, railbed of the Texas and New Orleans line. The lower deposits at the site had not been disturbed and have yielded a small, interesting collection of lithic implements.

Nine strata (Figure 20) were recognized in the excavations (Prewitt, 1966) and include (from top to bottom):

Zones IX-VII were sandy layers badly damaged by railroad construction and/or pothunters. Zones IX and VII were charcoal-strained and contained numerous burned rocks. Zone VIII was not present in all of the tested portions of the site, but where exposed it was found to be lighter in color and to contain only a few stone inclusions. None of these strata can be assigned to a time period.

Zone VI was a yellowish-tan, charcoal-strained layer which varied from 0.5 to 1.3 feet in thickness. It contained a fair amount of artifacts but the three points recovered cannot be assigned to a period.

Zone V, a light yellow, sandy clay, varied from 0.5 to 1.4 feet in thickness. The points from this layer are of a Period II style which compares quite favorably with the "Early Barbed" points found at the Devil's Mouth Site.

Zone IV was also a stratum of light yellow sandy clay. It yielded only one point, a specimen which closely resembles the "Early Barbed" form of Period II.

Zone III contained numerous small limestone spalls and was 0.5 to 1.6 feet thick. In the western part of the site, where the stratigraphy became quite complex, it merged with Zone Ie. The three points from Zone III cannot be typed but they are similar to the Gower-like point found in the Upper Gravels at the Devil's Mouth Site and the Bifurcated Stem points from Eagle Cave. Zone III is assigned to Period II.

Zone II, a light yellow sandy clay, reached a maximum thickness of two feet but lensed out in the western part of the site. Since the points from this

FIGURE 20.   Devils Rockshelter.   a, View looking southeast
             toward excavations.   b, East-west profile of
             strata uncovered in excavations.

a

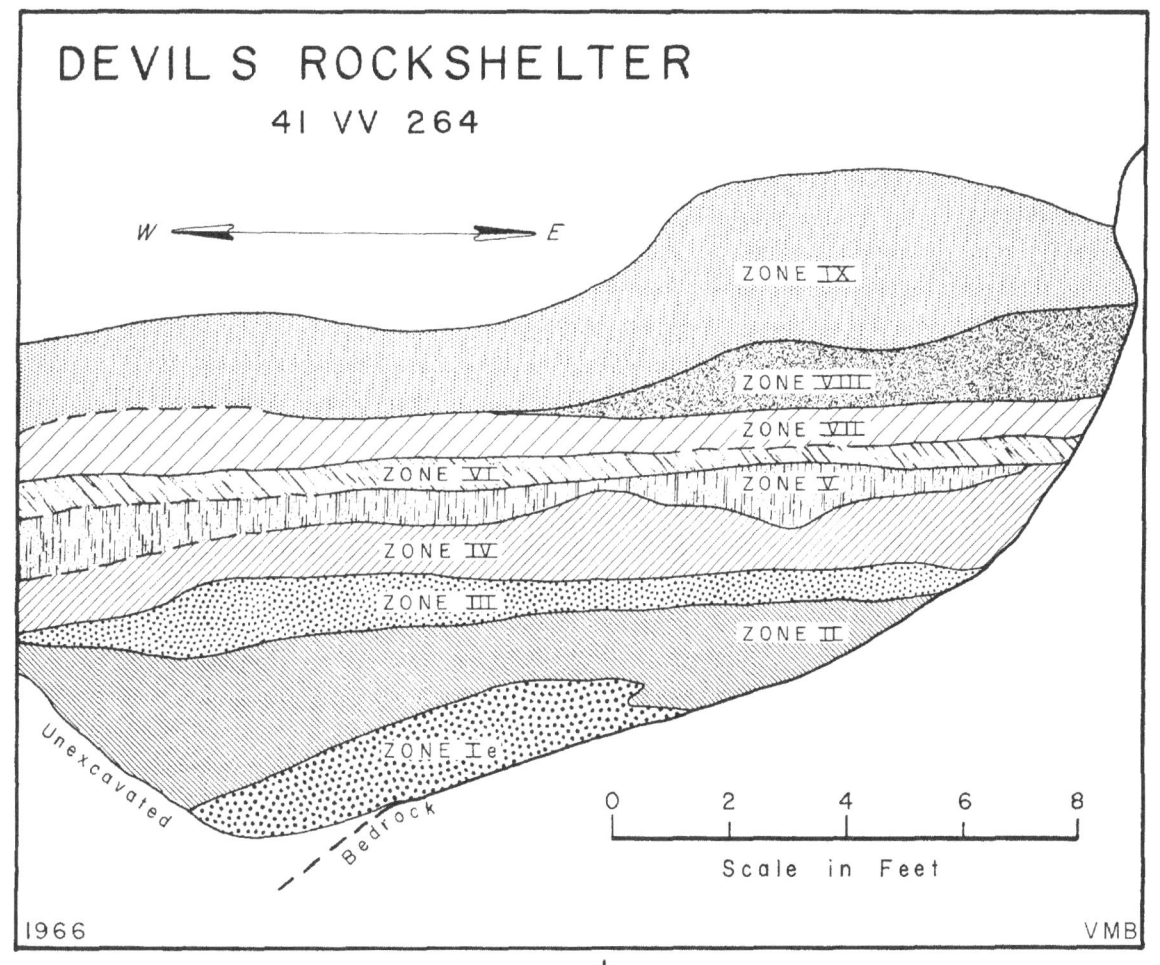

DEVILS ROCKSHELTER
41 VV 264

W ← → E

ZONE IX

ZONE VIII

ZONE VII

ZONE VI

ZONE V

ZONE IV

ZONE III

ZONE II

ZONE Ie

Unexcavated

Bedrock

0   2   4   6   8

Scale in Feet

1966   VMB

b

stratum are similar to those from Zone III, Zone II
is assigned to Period II.

Zone I, the lowest deposit encountered by the
T.A.S.P. excavations, was a complex layer which was
divided into five components (Ia, Ib, Ic, Id, and
Ie). It was clearly distinguished from Zone II, except in the western section of the site where the
stratigraphy became rather vague. Here Zone Ie (the
uppermost component)merged with Zone III. No points
were recovered from Zone Ia-Id, but two Period II
forms were found in the Zone Ie and III area.

INTRODUCTION TO BOTANICAL STUDIES

Donald A. Larson

The botanical studies conducted in conjunction with the work of archeologists in the Amistad region represent an attempt to provide archeologists with improved insights into primitive man's changing culture and the climates in which he lived. These projects - including 1) the collection of the modern flora of the region and a characterization of vegetation, 2) the identification of plant macrofossil from six well-documented archeological sites, and 3) pollen analyses of three shelter sites and one terrace site - are integrated among themselves and with the archeological studies. The preliminary reports are valuable to plant geographers, paleoclimatologists, palynologists, ecologists, and geomorphologists as well as to archeologists.

The collection of the plants of the region by Mr. David Flyr was included to provide a more complete understanding for the region's vegetation and for specimens useful in identification of plant macro- and microfossils. This collection was required because of the lack of previous collecting in this area both on the American and Mexican sides of the Rio Grande.

Macrofossils, initially, were looked upon as evidence of primitive man's usage of available plants. The primitive inhabitants of the area had already been characterized as hunters and gatherers rather than agriculturalists. The documentation of this characterization was sought. Only after the study was well underway did it become apparent that macrofossil evidence was of such a nature that it could be used to confirm and enhance microfossil evidence.

The magnitude of the macrofossil collections and the limitation of time and funds lead to a limitation of types of plant macrofossils to be identified. Wood fragments were excluded (but placed in storage for future study) because the time required was deemed too great for the expected information, especially since much of the wood would have been acquired as drift wood from any of the perennial rivers (Rio Grande, Devils, and Pecos). Fibrous artifacts were also excluded solely for reasons of time. The remaining plant materials consisted of fruits, seeds, flowers, plant fragments, and quids. As much of this material was identified as possible and the identifications are of representative portions of all habitation periods outlined by Story in the

31

introduction to the archeological section of this report.
Mr. Robert Irving has been responsible for the identifications
of all plant macrofossils.

Palynological studies were preceded by the preparation
of a pollen reference collection and a preliminary pollen
key for the area. Techniques of sample digestion were re-
fined to meet the peculiar requirements of the rockshelter
and terrace sediments encountered in this area. Four archeological sites were chosen for pollen analysis with the Devil's
Mouth Site and Bonfire Shelter receiving the most intensive
study. The decision to do so was based upon the obviously
greater potentialities of these sites. The other sites were
investigated to test whether stratigraphic correlations
among archeological sites in the area are possible. This
is considered important because certain sites or strata in
all sites do not yield material suitable for radiocarbon
dating. Also, the requirement of detecting differences in
pollen deposition between terrace and shelter sites was a
factor in the choice of Devil's Mouth Site and Bonfire Shelter.

THE CONTEMPORARY VEGETATION OF THE
AMISTAD RESERVOIR AREA

David Flyr

INTRODUCTION

This report covers observations and collections made
in the Amistad Reservoir area between October 4, 1965, and
April 24, 1966, on a total of eleven short trips.

The autumn in Val Verde County was warm and rather
dry, but good rains came during the winter and spring. With
so much moisture, an abundant spring flowering might have
been expected. Cold weather came late in the winter season and persisted through March, however, so that at the
end of the study period the vegetation was only approaching
the peak of spring flowering. Good collecting might have
been expected well into the summer of 1966.

The report includes first a general discussion of the
vegetation of the reservoir area, an attempt at recognizing
vegetational units, and finally a list of plants known so
far to occur in the area.

The author wishes to express his appreciation to
David Dibble and Mark Parsons for much assistance in the
field work and to Dr. Marshall C. Johnston of The University of Texas Herbarium for help in identification of the
specimens. All incorrect identifications and omissions are
the responsibility of the writer, however.

Collections made during the study period will be
deposited in the Herbarium of The University of Texas.

THE VEGETATION

The vegetation of Val Verde County is surprisingly
diverse considering the uniformity of the Cretaceous limestone substratum. The diversity may be explained by the
high relief created by the dissection of the substratum by
three major rivers and innumerable small and usually dry
streams. This dissection has created a number of habitats
differing greatly in the amount of moisture available to
plants. Development of soil in the area is generally
slight, there being a few areas of deep soil on the uplands
but mostly along the flood plain of the major streams.

From southeast to northwest there is a general rise
in elevation. The former area is essentially a continu-
ation of the South Texas Plains of Gould (1962), while the
latter area has many characteristics of the mesa country of
the Trans-Pecos region. In between, from north to south,
there is a gradual diminution of the typical Edwards
Plateau vegetation. The live oak-juniper vegetation so
typical of the Edwards Plateau is not found in the immediate vicinity of the reservoir. Live oak was found to
drop out some fifteen to twenty miles north of the reservoir area, while juniper was noted in scattered localities
within two or three miles of it. It is seen, then, that
the reservoir area cannot be comfortably included in any
of the major vegetation types usually assigned to Texas.

Four vegetational units are recognized in the reservoir area on the basis of study to date. With the exception of the first, the lines marking these units are not
sharp, but it has been thought best to make at least a
preliminary division as a basis for future work in the
area. Some conspicuous species are found in more than one
of the areas and tend to blur the lines between them.
Such plants will be noted in the discussion below.

Vega-Terrace Vegetation

This unit, occurring in the canyons of the Rio Grande
and Pecos as well as the lower Devils rivers, is characterized by a rather fine-textured sandy soil. Though the
vegetation occurs near permanent water, not all the species
are plants with a high water requirement. On the contrary,
plants occurring on the upper portions of the terrace have
little available water, since rainfall passes rapidly
through the upper layers of the fine sand and is lost to
plants growing there.

Plant growth here is often quite luxuriant. The most
striking aspect of the unit is the fact that most of the
major species occurring here are not native but have been
introduced, some accidentally, from South America or the
Old World. Along the Rio Grande, much of the ground is
covered with a heavy growth of one of these introductions;
Bermuda grass (Cynodon dactylon). Also abundant is the
giant reed (Arundo donax). Yet another introduction, the

common reed (<u>Phragmites</u> <u>communis</u>), is frequent in the lower Pecos canyon and very abundant on the rocky bed of the lower Devils River. Of sporadic occurrence is the old world salt cedar (<u>Tamarix</u> <u>gallica</u>), a plant which completely dominates the valley of the Pecos in the upper portion of its Texas drainage. Also common among the introduced species are tree tobacco (<u>Nicotiana</u> <u>glauca</u>) and castor bean (<u>Ricinus</u> <u>communis</u>).

Large trees in this unit are rare. On the Pecos, this might be the result of the major flood of 1954 when water reached a depth of 30 meters in the river canyon. Mesquite (<u>Prosopis</u>) does occur as a small tree, and there are scattered willows (<u>Salix</u>) on the Pecos and Rio Grande banks. Near the mouth of the Devils River, there are small stands of willow, sycamore (<u>Platanus</u>), pecan (<u>Carya</u>), and mulberry (<u>Morus</u>).

On the higher terraces, two species of cactus, <u>Opuntia</u> <u>lindheimeri</u> and <u>O.</u> <u>leptocaulis</u> sometimes occur, along with several herbaceous species including bitterweed (<u>Hymenoxys</u>) and crownbeard (<u>Ximenesia</u>). Forming dense clumps at various places in this unit, often up to three meters in height, is the globe mallow (<u>Sphaeralcea</u> <u>angustifolia</u>). This native and two others, the seepwill baccharis (<u>Baccharis</u> <u>glutinosa</u>) and devil-weed aster (<u>Aster</u> <u>spinosus</u>), are the only native species found in abundance in this unit.

As a whole, then, the vegetation of the vegas and terraces is rich in quantity but poor in number of species. It is of a weedy nature, reflecting the disturbing factors of flowing water, flood and drought.

Cliff-Canyon Vegetation

This unit, though not always well differentiated from those to follow, is quite distinct from the vegetation of the river terraces. Mesquite, for example, occurs frequently on the terraces but stops abruptly where the sandy terrace soil gives way to the rocky thin (if any) soil of the small side canyons leading into the major streams. Most of the canyons are narrow and quite steep. The central portions of them are occupied by large shrubs or small trees, these creating so dense a shade that herbaceous vegetation is often limited to the more gently sloping margins.

Here are found the western soapberry (Sapindus drummondii) and the Mexican buckeye (Ungnadia speciosa), along with two species of hackberry (Celtis laevigata and C. reticulata -- the former in more mesic situations than the latter), the little walnut (Juglans microcarpa), and the vasey shin oak (Quercus pungens var. vaseyana). (Other species of oak no doubt occur in the canyons, but only one was collected.) The rare Texas pistacia (Pistacis texana) was found in two of the several canyons visited. Other minor components of this unit include Bumelia, wafer ash (Ptelea), and leadtree (Leucaena). The Mexican ash (Fraxinus berlandieriana) is rather rare but may become quite a large tree in the moister parts of the canyons, while the common gregg ash (F. greggii) is usually no more than a large shrub and occupies the dry upper portions of the canyons, tending to merge with the upland vegetation. Other shrubs which occupy both canyons and surrounding hills are guajillo (Acacia berlandieri), Texas colubrina (Colubrina texensis), coyotillo (Karwinskia humboldtiana) and mescal bean (Sophora secundiflora). The genus Ephedra was found mostly in rocky cliffs, but this may be the result of its having been eliminated from other areas by grazing sheep and goats.

A very interesting feature of this unit and one which deserves much further study is the group of species which grow in the vertical fissures of the limestone canyon walls. These include the rare thistle Cirsium turneri, the rayless rock daisy (Perityle angustifolia), the baccharis leaf penstemon (Penstemon baccharifolius), and a number of other species. The first of these was found growing in vertical walls of only one canyon leading into the Pecos River.

Finally, several species are restricted to moist, shady areas of the canyons and might be considered indicators of mesic conditions. Such species are the rouge plant (Rivina humilis), snapdragon maurandya (Maurandya antirrhinifolia), and yellow rock-nettle (Eucnide bartonioides).

Upland Vegetation (Hills)

Division of the upland vegetation into two units is somewhat arbitrary, yet the extremes are diverse enough to warrant this recognition even if there is intergradation between the two.

The rocky upland hills in the reservoir area demonstrate a transition between quite different vegetational types. As has been noted, the typical Edwards Plateau vegetation is not found in the immediate vicinity. Only scattered junipers (_Juniperus ashei_) are found and no live oaks (_Quercus virginiana_ var. _fusiformis_); other common species of the Plateau such as evergreen sumac (_Rhus virens_) and agarita (_Berberis trifoliolata_) are seen only occasionally and then in or near small canyons. Mescal bean (_Sophora secundiflora_), common on the uplands north of the Amistad area, is a typical element of the canyon vegetation already described. No specimens of the Edwards Plateau species flameleaf sumac (_Rhus copallina_ var. _lanceolata_) were found.

The existence of an apparently relic population of Mexican pinyon (_Pinus cembroides_) about 30 miles north of Del Rio is remarkable, indicating that the area is drier now than in the recent past since the pinyon is characteristic of the moister middle elevation in Trans-Pecos Texas.

The transition mentioned above takes place mostly in an east-west direction. To the east the hills are covered with guajillo (_Acacia berlandieri_) and blackbrush acacia (_A. rigidula_). Ceniza (_Leucophyllum frutescens_) is common here but becomes the dominant shrub further to the west in the Comstock area. Guayacan (_Porlieria angustifolia_) and brush myrtlecroton (_Bernardia myricaefolia_) also increase in abundance in the western portion of the area. One of the most important plants in much of the reservoir area is the Texas persimmon (_Diospyros texana_). This large shrub often produces an abundance of fruit and may have been important in the diet of early inhabitants of the region. Near the Devils River and continuing westward, the gregg ash (_Fraxinus greggii_) is common on the hills near small canyons. At the same place, the first specimens of ocotillo (_Fouquieria splendens_) are seen. This plant is definitely one of the distinct indications of the beginning of a true desert flora like that of Trans-Pecos Texas. _Agave lecheguilla_ is another of the same type, and its eastern boundary seems also to lie near the Devils River. Still another typical western species is the skeleton golden-eye (_Viguiera stenoloba_). Though not restricted as the above, species of _Yucca_, _Nolina_ and _Dasylirion_ first begin to be prominent west of the Devils River but east of the Pecos.

Thus the character of the upland vegetation changes from that seen on parts of the South Texas Plains as described by Johnston (1955) to that of the Trans-Pecos country described by Tharp (1944), Webster (1950) and Warnock (1946).

Upland Vegetation (Flats)

An east-west transition is also seen in this unit, which includes areas with some soil, usually fine-textured, lying between the rocky limestone hills. The most abundant species; mesquite (_Prosopis glandulosa_) and spiny hackberry (_Celtis pallida_) are of common occurrence on the South Texas Plains. Other common species typical of South Texas are the condalias, _Zixyphus_, berlandier wolfberry (_Lycium berlandieri_), tasajillo (_Opuntia leptocaulis_) and shrubby blue sage (_Salvia ballotaeflora_). Beyond the Devils River, one sees for the first time species which occupy vast areas of desert flat-lands to the west: creosote bush (_Larrea divaricata_) and tarbush (_Flourensia cernua_).

Herbaceous vegetation on the uplands is rather sparse, at least partly the result of overgrazing of livestock. Some three species constitute the major portion of the herbaceous element. They are the common dogweed (_Dyssodia pentachaeta_), grassland croton (_Croton dioicus_) and one grass, tobosa (_Hilaria mutica_). It is quite possible that on the Mexican side of the Rio Grande where grazing may have been less severe, more herbaceous species might be found.

SUMMARY

Certain points may now be made regarding the reservoir area as a whole. Though the region is seen to bear relationship to the South Texas Plains, the Edwards Plateau, and to the deserts of the Trans-Pecos, it is still distinct from all these in one or more respects. The relations are much stronger with the Mexican (even South American) deserts and sub-tropics than with the typical vegetation of any area of the United States.

Among the Compositae, the largest family in numbers of
species, only a single species of the tubiflorous subfamily
is found: <u>Pinaropappus</u> <u>roseus</u>. A single species of
<u>Artemisia</u> was found and this of very minor importance, a
situation quite different from that obtaining in the western
United States.

Species of Chenopodiaceae and Amaranthaceae were also
very minor components of the flora.

Bray (1905) speaks of the distinctness of what he calls
the Sotol Country as follows: "...one gains the impression
that this southwestern plateau desert has had a more effective
barrier about it then has the continent as a whole...."

Much more collection and study should be done in this
area. Many other species may be found which will link the
Amistad area to other floristic provinces to the east, south
or west. At present, we can regard it has a rather distinct
region with relations to all of the other three but not to be
considered a part of any.

THE PLANT LIST

The following list of plants represents collections
made in the study area along with a limited number of
species known from previous collections to occur there but
not seen by the author during the study period. A few of
the species were observed but not collected. The nomenclature follows that of Gould's (1962) checklist except in
those cases where taxonomic revision of certain groups has
appeared since the publication of the checklist. Valid
differences of opinion may exist regarding the nomenclature
used in the checklist, but it has been thought best to follow
Gould closely to make this report of maximum effectiveness,
since no more complete account of the flora of Texas exists
at present.

Certain major deficiencies in the list following may
be noted. First, very few species of grasses appear. This
may be explained by the fact that few grass species are
conspicuous or important elements of the contemporary flora.
The Cactaceae are poorly represented because the time of

study did not fall within the major flowering period of that family, and identification of sterile material is a hazardous undertaking. With the exception of two species of <u>Yucca</u>, the liliaceous general <u>Yucca</u>, <u>Nolina</u>, and <u>Dasylirion</u> also did not flower during the study period, and identification in these genera is tentative. Finally, identification of species of <u>Ephedra</u> is uncertain. Those listed below have been collected in the study area and were cited in the last monograph of that genus.

A PRELIMINARY LIST OF PLANTS KNOWN TO

OCCUR IN THE AMISTAD RESERVOIR AREA

(Plants of special interest included in the discussion above are marked by an asterisk.)

EQUISETACEAE (Horsetail Family)

Equisetum laevigatum A. Br. Smooth Horsetail

SELAGINELLACEAE (Selaginella Family)

Selaginella lepidophylla Hook. & Grev. Resurrection Plant

Selaginella wrightii Hieron. Wright Selaginella

POLYPODIACEAE (Fern Family)

Cheilanthes aemula Maxon Lipfern

Cheilanthes tomentosa Link Woolly Lipfern

Notholaena copelandii Hall Cloakfern

Pellaea ovata (Desv.) Weatherby Cliffbrake

PINACEAE (Pine Family)

Juniperus ashei Buchholz Ashe Juniper

Pinus cembroides Zucc. Mexican Pinyon

EPHEDRACEAE (Joint-fir Family)

Ephedra antisyphlitica Berl. ex C. A. Meyer

Ephedra nevadensis Wats. var. aspera (Engelm.)

L. Benson Boundary Ephedra

Ephedra pedunculata Engelm.

GRAMINEAE (Grass Family)

Arundo donax L. Giant Reed

Boutelous trifida Thurb. Red Grama

Cynodon dactylon (L.) Pers. Bermudagrass

Erioneuron pilosum (Buckl.) Nash

Hilaria mutica (Buckl.) Benth. Tobosa

Phragmites communis Trin. Common Reed

Sporobolus wrightii Munro Big Sacaton

Tridens muticus (Torr.) Nash Slim Tridens

CYPERACEAE (Sedge Family)

Carex brittoniana Bailey Britton Sedge

Eleocharis montevidensis Kunth Sand Spikesedge

Scirpus acutus Muhl. Hardstem Bullrush

Scirpus olneyi Gray Olney Bullrush

LILIACEAE (Lily Family)

Allium drummondi   Regel   Drummond Onion

Dasylirion texanum   Scheele   Texas Sotol

Hesperaloe parviflora   (Torr.)   Coulter   Red Hesperaloe

Nolina texana   S. Wats.   Sacahuiste

Smilax bona-nox   L.   Saw Greenbrier

Yucca constricta   Buckl.   Buckley Yucca

Yucca thompsoniana   Trel.   Thompson Yucca

Yucca torreyi   Shafer   Torrey Yucca; Spanish Dagger

AMARYLLIDACEAE (Amaryllis Family)

Agave lecheguilla   Torr.   Lechuguilla

SALICACEAE (Willow Family)

Salix interior   Rowlee   Sandbar Willow

Salix nigra   Marsh   Black Willow

JUGLANDACEAE (Walnut Family)

Carya illinoensis   (Wang.) K. Koch   Pecan

Juglans microcarpa   Berland.   Little Walnut

FAGACEAE (Oak Family)

Quercus *pungens* Liebm. var. *vaseyana* (Buckl.)

C. H. Muller Vasey Shin Oak

Quercus *virginiana* Mill. var. *fusiformis* (Small)

Sarg. Plateau Oak

ULMACEAE (Elm Family)

Celtis *laevigata* Willd. Sugar Hackberry

Celtis *pallida* Torr. Spiny Hackberry

Celtis *reticulata* Torr. Netleaf Hackberry

MORACEAE (Mulberry Family)

Morus *alba* L. White Mulberry

Morus *microphylla* Buckl. Texas Mulberry

URTICACEAE (Nettle Family)

Parietaria *pennsylvanica* Muhl. var. *obtusa* (Rydb.)

Shinners Pennsylvania Pellitory

Urtica *chamaedryoides* Pursh Heartleaf Nettle

LORANTHACEAE (Mistletoe Family)

Phoradendron *serotinum* (Raf.) M. C. Johnston var.

pubescens (Engelm.) M. C. Johnston Mistletoe

POLYGONACEAE (Buckwheat Family)

Eriogonum longifolium Nutt. Longleaf Wildbuckwheat

CHENOPODIACEAE (Goosefoot Family)

Chenopodium berlandieri Moq. Pitseed Goosefoot

Salsola kali L. Russian Thistle; Tumbleweed

NYCTAGINACEAE (Four-O'Clock Family)

Acleisanthes longiflora Gray Angel Trumpets

Allionia incarnata L. Trailing Allionia

Boerhaavia linearifolia Gray Narrowleaf Spiderling

Cyphomeris gypsophiloides (Mart. & Gal.) Standl.
   Red Cyphomeris

Mirabilis grayana (Standl.) Standl. Gray Four-o'clock

PHYTOLACCACEAE (Pokeweed Family)

Rivina humilis L. Bloodberry Rougeplant

PORTULACACEAE (Purslane Family)

Portulaca pilosa L. Shaggy Purslane

CARYOPHYLLACEAE (Pink Family)

Paronychia jamesii T. & G. James Nailwort

RANUNCULACEAE (Crowfoot Family)

Clematis drummondii T. & G.   Texas Virgins Bower

BERBERIDACEAE (Barberry Family)

Berberis trifoliolata Moric. Agarito

PAPAVERACEAE (Poppy Family)

Corydalis aurea Willd. var. occidentalis Engelm.

    Golden Corydalis

CRUCIFERAE (Mustard Family)

Descurainia pinnata (Walt.) Britt. Tansymustard

Draba cuneifolia Nutt. Wedgeleaf Draba; Whitlow Grass

Lepidium austrinum Small Southern Pepperweed

Lepidium virginicum L. Virginia Pepperweed

Lesquerella fendleri (Gray) Wats. Fendler Bladderpod

Lesquerella gordoni (Gray) Wats. Gordon Bladderpod

Lesquerella purpurea (Gray) Wats. Rose Bladderpod

Sibara runcinata (Wats.) Rollins

Streptanthus platycarpus Gray Broadpod Twistflower

CAPPARIDACEAE (Caper Family)

Polanisia dodecandra (L.) DC. var. trachysperma (T. & G.)
Iltis   Roughseed   Clammyweed

CRASSULACEAE (Stonecrop Family)

Sedum sp.   Stonecrop

PLATANACEAE (Sycamore Family)

Platanus occidentalis   L.   American Sycamore

ROSACEAE (Rose Family)

Prunus minutiflora   Engelm.   Smallflower Peachbrush

Prunus serotina   Ehrhart subsp. virens   (Woot. & Standl.)
McVaugh   Southwestern Chokecherry

Rubus trivialis   Michx.   Southern Dewberry

LEGUMINOSAE (Pea Family)

Acacia berlandieri   Benth.   Guajillo

Acacia farnesiana   (L.)   Willd.   Huisache

Acacia greggii   Gray   Catclaw Acacia

Acacia rigidula   Benth.   Blackbrush Acacia

Acacia roemeriana   Scheele   Roemer Acacia

Acacia vernicosa   Standl.   Stickyleaf Acacia

Calliandra conferta Benth. ex Gray

Cassia lindheimeriana Scheele Lindheimer Senna

Cassia roemeriana Scheele Twoleaf Senna

Cercidium texanum Gray Texas Paloverde

Dalea formosa Torr. Feather Dalea

Dalea frutescens Gray var. laxa (Rydb.) Turner Black
    Dalea

Dalea nana Torr. ex Gray var. elatior Gray Dwarf Dalea

Dalea pogonathera Gray var. walkerae (Tharp & Barkley)
    Turner Bearded Dalea

Eysenhardtia texana Scheele Texas Kidneywood

Krameria grayi Rose & Painter White Ratany

Krameria lanceolata Torr. Trailing Ratany

Leucaena retusa Benth. ex Gray Littleleaf Leadtree

Mimosa biuncifera Benth. var. biuncifera Catclaw Mimosa
    var. lindheimeri (Gray) Robinson Lindheimer Mimosa

Prosopis glandulosa Torr. var. glandulosa Honey Mesquite

Sophora secundiflora (Ort.) Lag. ex DC. Mescalbean;
    Texas Mountainlaurel

Vicia leavenworthii Torr. & Gray var. occidentalis
    Shinners Leavenworth Vetch

OXALIDACEAE (Sorrel Family)

Oxalis dichondraefolia Gray Ponyleaf Woodsorrel

LINACEAE (Flax Family)

Linum rupestre Engelm. Rock Flax

ZYGOPHYLLACEAE (Caltrop Family)

Larrea divaricata Cav. Creosote Bush

Porlieria angustifolia (Engelm.) Gray Guayacan

RUTACEAE (Rue Family)

Ptelea trifoliata   L. Hoptree; Wafer Ash

Thamnosma texana (Gray) Torr. Texas Desertrue

SIMAROUBACEAE

Castela texana (T. & G.) Rose Allthorn Goatbush

POLYGALACEAE (Milkwort Family)

Polygala lindheimeri Gray Shrubby Milkwort

Polygala ovalifolia Gray Eggleaf Milkwort

EUPHORBIACEAE (Spurge Family)

Acalypha hederacea Torr. Copperleaf

Acalypha lindheimeri Muell. Arg. Lindheimer Copperleaf

Argythamnia neomexicana Muell. Arg. New Mexico Wildmercury

Bernardia myricaefolia (Scheele) Wats. Brush Myrtlecroton

Bernardia obovata I. M. Johnston Desert Myrtlecroton

Croton dioicus Cav. Grassland Croton

Croton fruticulosus Engelm. Bush Croton

Croton torreyanus Muell. Arg. Torrey Croton

Euphorbia cinerascens Engelm. Ashy Euphorbia

Euphorbia glyptosperma Engelm. Ridgeseed Euphorbia

Euphorbia spathulata Lam. Warty Euphorbia

Jatropha dioica Sesse ex Cerv. var. dioica Leatherstem

Phyllanthus polygonoides Nutt. ex Spreng. Knotweed

Leafflower

Ricinus communis L. Castorbean

Tragia nepetaefolia Cav. Catnip Noseburn

ANACARDIACEAE (Cashew Family)

Pistacia texana Swingle Texas Pistachia

Rhus microphylla Engelm. ex Gray Littleleaf Sumac

Rhus radicans L. Poison Ivy

Rhus virens Lindh. ex Gray Evergreen Sumac

CELASTRACEAE (Bittersweet Family)

Schaefferia cuneifolia Gray Capul

SAPINDACEAE (Soapberry Family)

Sapindus drummondii Hook. & Arn. Western Soapberry

Ungnadia speciosa Endl. Mexican Buckeye

RHAMNACEAE (Buckthorn Family)

Colubrina texensis (T. & G.) Gray Texas Colubrina; Hog-plum

Condalia hookeri M. C. Johnston

Condalia viridis I. M. Johnston Green Condalia

Karwinskia humboldtiana (R. & S.) Zucc. Coyotillo

Zizyphus obtusifolia (Hook. ex T. & G.) Gray Southwest
   Condalia; Lotebush

VITACEAE (Grape Family)

Cissus incisa (Nutt.) Des Moulins Ivy Treebine

Vitis arizonica Englem. Canyon Grape

Vitis cinerea Engelm. Sweet Grape

MALVACEAE (Mallow Family)

Abutilon incanum (Link.) Sweet Indianmallow

Abutilon wrightii Gray Wright Abutilon

Hibiscus cardiophyllus Gray Heartleaf Rosemallow

Hibiscus coulteri Gray Desert Rosemallow

Sida filicaulis T. & G. Spreading Sida

Sida filipes Gray Violet Sida

Sphaeralcea angustifolia (Cav.) D. Don var. angustifolia
     Narrowleaf Globemallow

Sphaeralcea hastatula Gray Spear Globemallow

STERCULIACEAE (Cacao Family)

Melochia pyramidata L. Anglepod Melochia

TAMARICACEAE (Tamarisk Family)

Tamarix gallica L. Saltcedar

FOUQUIERIACEAE (Ocotillo Family)

Fouquieria splendens Engelm. Ocotillo

KOEBERLINIACEAE (Junco Family)

Koeberlinia spinosa Zucc. Allthorn, Crucifixion Thorn

PASSIFLORACEAE (Passion Flower Family)

Passiflora tenuiloba Engelm. Spreadlobe Passionflower

LOASACEAE (Loasa Family)

Cevallia sinuata Lag. Stinging Cevallia

Eucnide bartonioides Zucc. Yellow Rocknettle

Mentzelia lindheimeri Urban & Gilg. Lindheimer
      Mentzelia; Stickleaf

Mentzelia pumila (Nutt.) T. & G. Yellow Mentzelia

CACTACEAE (Cactus Family)

Mammillaria heyderi Muhlenpfordt Heyder Mammillaria

Opuntia leptocaulis DC. Pencil Cactus; Tasajillo

Opuntia lindheimeri Engelm. Texas Pricklypear

Opuntia phaeacantha Engelm. & Bigel. var. major Engelm.
      Brownspine Pricklypear

ONAGRACEAE (Evening-Primrose Family)

Oenothera greggii Gray Gregg Eveningprimrose

Oenothera hartwegii Benth. var. hartwegii Hartweg Primrose

Oenothera lavendulaefolia Torr. & Gray Lavenderleaf
      Eveningprimrose

Oenothera serrulata Nutt. Halfshrub Sundrops

UMBELLIFERAE (Parsley Family)

Bowlesia incana Ruiz & Pav. Hoary Bowlesia

54

PRIMULACEAE (Primrose Family)

Samolus ebracteatus HBK. subsp. *cuneatus* (Small)
 R. Kunth Limerock Brookweed

SAPOTACEAE

Bumelia lanuginosa (Michx.) Pers. var. *texana* (Buckl.)
 Cronquist

EBENACEAE (Ebony Family)

Diospyros texana Scheele Texas Persimmon

OLEACEAE (Olive Family)

Forestiera pubescens Nutt. Elbowbush

Fraxinus berlandieriana A. DC. Mexican Ash

Fraxinus greggii Gray Gregg Ash

Menodora longiflora Gray Showy Menodora

GENTIANACEAE (Gentian Family)

Centaurium calycosum (Buckl.) Fern. Buckley Centaury

Centaurium texense (T. & G.) A. DC. Texas Centaury

APOCYNACEAE (Dogbane Family)

Macrosiphonia macrosiphon (Torr.) Heller Plateau Rocktrumpet

ASCLEPIADACEAE (Milkweed Family)

Cynanchum barbigerum (Scheele) Shinners var. barbigerum
  Bearded Swallowwort

Cynanchum maccartii Shinners McCart Swallowwort

Matelea woodsonii Shinners Woodson Milkvine

CONVOLVULACEAE (Morningglory Family)

Cuscuta glabrior (Englem.) Yuncker Dodder

Cuscuta indecora Choisy Showy Dodder

Evolvulus alsinoides L. Slender Evolvulus

Ipomoea lindheimeri Gray Blue Morningglory

Merremia dissecta (Jacq.) Hallier f.

POLEMONIACEAE (Phlox Family)

Gilia incisa Benth. Splitleaf Gilia

Gilia rigidula Benth. Prickleaf Gilia

HYDROPHYLLACEAE (Waterleaf Family)

Phacelia congesta Hook. Spike Phacelia

Phacelia patuliflora Gray Sand Phacelia

BORAGINACEAE (Borage Family)

Coldenia canescens DC. Gray Coldenia

Heliotropium curassavicum L. Salt Heliotrope

Heliotropium torreyi I. M. Johnston Slimleaf Heliotrope

Lappula redowskii (Hornem.) Greene Flatspine Stickseed

Omphalodes aliena Gray Mexican Navelseed

VERBENACEAE (Verbena Family)

Aloysia lycioides Cham. var. schulzii (Standl.)
    Moldenke Whitebrush

Lippia graveolens HBK. Scented Lippia

Tetraclea coulteri Gray

Verbena pumila Rydb. in Small Pink Vervain

LABIATAE (Mint Family)

Salvia ballotaeflora Benth. Shrubby Blue Sage

Salvia farinacea Benth. Mealycup Sage

Salvia texana (Scheele) Torr.

SOLANACEAE (Nightshade Family)

Lycium berlandieri Dunal Berlandier Wolfberry

Nicotiana glauca Graham Tree Tobacco

Nicotiana trigonophylla Dunal Desert Tobacco

Nierembergia viscosa Torr. Texas Cupflower

Petunia parviflora Juss. Wild Petunia

Physalis lobata Torr. Purple Groundcherry

<u>Physalis</u> <u>viscosa</u> L. var. <u>cinerascenus</u> (Dunal)

   Waterfall Groundcherry

<u>Solanum</u> <u>americanum</u> Mull. Blueflower Buffalobur

<u>Solanum</u> <u>triquetrum</u> Cav. Texas Nightshade

      SCROPHULARIACEAE (Figwort Family)

<u>Castilleja</u> <u>latebracteata</u> Pennell Broadbract Paintbrush

<u>Leucophyllum</u> <u>frutescens</u> (Berl.) I. M. Johnston Ceniza

<u>Maurandya</u> <u>antirrhiniflora</u> Humb. & Bonpl. Snapdragon

   Maurandya

<u>Penstemon</u> <u>baccharifolius</u> Hook. Baccharisleaf Penstemon

      BIGNONIACEAE (Bignonia Family)

<u>Chilopsis</u> <u>linearis</u> (Cav.) DC. Desertwillow

     OROBANCHACEAE (Broomrape Family)

<u>Orobranche</u> <u>ludoviciana</u> Nutt. Louisiana Broomrape

     ACANTHACEAE (Acanthus Family)

<u>Ruellia</u> <u>parryi</u> Gray Parry Ruellia

<u>Siphonoglossa</u> <u>pilosella</u> (Nees) Torr. Hairy Tubetongue

PLANTAGINACEAE (Plantain Family)

Plantago rhodosperma Dcne. Redseed Plantain

RUBIACEAE (Madder Family)

Hedyotis acerosa Gray Needleleaf Bluets

Hedyotis nigricans (Lam.) Fosberg Prairie Bluet

CUCURBITACEAE (Gourd Family)

Ibervillea tenuisecta (Gray) Small Slimlobe Globeberry; Deer-apples

Melothria pendula L. Melonette

CAMPANULACEAE (Bluebell Family)

Triodanis coloradoensis (Buckl.) McVaugh Colorado Venuslookingglass

COMPOSITAE (Sunflower Family)

Aphanostephus riddellii T. & G. Riddell Dozedaisy

Artemisia ludoviciana Nutt. Louisiana Sagewort

Aster spinosus Benth. Devilweed Aster

Baccharis glutinosa Pers. Seepwillow Baccharis

Baccharis salicina T. & G. Willow Baccharis

Bahia absinthifolia Benth. Hairyseed Bahia

Brickellia laciniata Gray Splitleaf Brickellbush

Chaetopappa bellidifolia (Gray & Engelm.) Shinners
      Hairy Leastdaisy

Chaetopappa bellioides (Gray) Shinners Manyflower Least-
      daisy

Cirsium turneri Warnock Turner Thistle

Dyssodia acerosa DC. Prickleaf Dogweed

Dyssodia micropoides (DC.) Loesener Woolly Dogweed

Dyssodia pentachaeta (DC.) Robinson Common Dogweed

Dyssodia tenuiloba (DC.) Robinson Bristleleaf Dogweed

Erigeron modestus Gray Plains Fleabane

Eupatorium greggii Gray Palmleaf Eupatorium

Flourensia cernua DC. Tarbush

Grindelia grandiflora Hook. Manyray Gumweed

Heterotheca villosa (Pursh) Shinners Hairy Goldaster

Hymenoclea monogyra T. & G. Burrobrush

Hymenoxys odorata DC. Western Bitterweed

Liatris mucronata DC. Gayfeather

Machaeranthera australis (Greene) Shinners

Melampodium leucanthum T. & G. Plains Blackfoot

Palafoxia callosa (Nutt.) T. & G. var. _bella_
    (Cory) Shinners Small Palafoxia

Palafoxia texana DC. var. _texana_ Texas Palafoxia

Parthenium confertum Gray var. microcephalum Rollins

Parthenium hysterophorus L. Ragweed Parthenium

Perezia runcinata (D. Don) Gray Stemless Perezia

Perityle angustifolia (Gray) Shinners Rayless Rockdaisy

Pinaropappus roseus Less. White Rocklettuce

Porophyllum scoparium Gray Poreleaf

Psilostrophe gnaphaloides DC. Cudweed Paperflower

Psilostrophe villosa Rydb. Hairy Paperflower

Senecio longilobus Benth. Threadleaf Groundsel

Simsia calva (Engelm. & Gray) Gray Awnless Bushsunflower

Tetragonotheca texana (Gray) Engelm. & Gray Plateau
     Nerveray

Tetraneuris scaposa (DC.) Greene Plains Tetraneuris

Thelesperma longipes Gray Longstalk Greenthread

Thelesperma megapotamicum (Spreng.) Kuntze Colorado
     Greenthread

Thelesperma simplicifolium Gray Slender Greenthread

Viguiera stenoloba Blake Skeleton Goldeneye

Xanthocephalum sarothrae (Pursh) Shinners Broom
     Snakeweed

Xanthocephalum texanum (DC.) Shinners Texas Broomweed

Ximenesia encelioides Cav. Golden Crownbeard

Zexmenia hispida (HBK.) Gray

A PRELIMINARY ANALYSIS OF PLANT REMAINS FROM
SIX AMISTAD RESERVOIR SITES

Robert S. Irving

INTRODUCTION

The identification of collected plant macrofossils was undertaken with the intention of gaining an insight into primitive man's diet and plant usage, and to gain experience with the problems of macrofossil collection and identification so as to expedite future work in this area of study. During the initial stages of this project the author organized a reference collection of plants and plant parts for comparative purposes. This collection is currently being expanded and is supplemented by the seed identification collection of the Herbarium of The University of Texas.

One benefit of the macrofossil report which follows is the degree in which it integrates with the pollen record in demonstrating changes in vegetational patterns. However, in this respect care must be taken to insure that this is not the results of changing cultural patterns rather than vegetational changes that are being detected.

Identification of the total collection of plant parts and fragments was deemed impossible because of time limitations, therefore, randomly selected samples of the seed, fruit, flower, and fibrous material from archeological strata in six sites in the Amistad Reservoir were studied. Wood fragments and vegetable material used in the manufacture of artifacts have not been identified as yet. However, these have been set aside for future identification.

Identification of the specimens from the several sites are listed and the frequency of occurrence (actually the number of lot or cataloging numbers) of each taxonomic entity is noted in tabular form. The organization of the results conforms to occupational periods in descending order as outlined by Story (see introduction to this report).

METHODS

The primary method employed in this identification work was that of comparison to knowns. A reference collection was assembled of the known, more commonly occurring, plants of the area, as determined from Gould (1962), past identification work, and The University of Texas Herbarium distribution records. This collection consisted of floral parts, fruits, seeds, and vegetational structures. Each aspect was placed in a separate paper packet which was in turn filed alphabetically by genus. Further, representatives of the more commonly occurring variation patterns of these parts were also included.

As identification proceeded the collection was expanded to include the newly encountered species and genera. Also, several representatives of the plant megafossils themselves (where they were abundant) were included in the appropriate packets. This was done in order to have a reference to the variation (especially in color) due to aging and decay.

In all cases where uncertainty existed herbarium specimens were utilized to the fullest as an extension of the reference collections.

The names applied to the megafossils represent, wherever possible, that of the latest taxonomic treatment. Where no such treatment existed the name was chosen that occurs in Gould's 1962 checklist. In this application of names it must be noted that there is some chance of error. This results from the fossils in some species only reflecting a portion of the characters needed for definite name assignment. In most cases, however, the morphology of the fossil or morphology coupled with distribution data did allow for the giving of a specific name. In those cases where such uncertainty existed the specific name was omitted.

In the handling of the material it became necessary to make some reassortments. In many cases (however, not all) bags containing heterogeneous samples were divided into component taxa by the use of paper packets which were rebagged in the initial container. Further, all nonidentified vegetative material, including charcoal, decayed wood and twigs, were segregated from the identified samples and stored separately. This was also done for non-vegetative material encountered. In all cases where there was separation, either non-vegetative or non-identifiable, the samples were relabeled with the appropriate data.

ALPHABETICAL LISTING OF SPECIFIC AND COMMON NAMES
OF IDENTIFIED PLANTS*

Acacia berlandieri Benth. (guajillo)

Acacia greggii Gray (cat's claw, devil's claw, chaparral)

Acacia rigidula Benth. (blackbush acacia, blackbush)

Acacia roemeriana Scheele (roemer acacia)

Agave lecheguilla Torr. (lechuguilla)

Agave sp. (agave)

Allium drummondi Regel (drummond onion)

Ariocarpus fissuratus [Engelm.] Schumann (chautle living
rock)

Ariocarpus sp.

Aristida glauca [Nees] Walp. (blue three-awn, reverchon
three-awn)

Carya sp. (hickory)

Cassia roemeriana Scheele (two leaf senna)

Celtis reticulata Torr. (netleaf hackberry)

Cucurbita foetidissima H.B.K. (buffalo gourd, stinking
gourd, calabazilla,
Missouri gourd, fetid
wild pumpkin)

Diospyros texana Scheele (Texas persimmon)

Helianthus sp. (sunflower)

Jatropha dioica Sesse ex Cerv. (leather stem, sangre de
drago, rubber plant)

Juglans microcarpa Berland (little walnut, Texas black
walnut)

Juglans sp. (walnut)

---

*Spelling and common names after Gould, 1962.

Karwinskia humboldtiana [R.S.S.] Zucc. (coyotillo)

Leucaena retusa Beth. ex Gray (little-leaf leadtree)

Opuntia lindheimeri Engelm. (Texas prickly pear, nopal prickly pear)

Opuntia phaeacantha Engelm. & Bigel. (brownspine prickly pear)

Opuntia sp. (prickly pear)

Pappophorum bicolor Fourn. (two-colored pappusgrass)

Prosopis grandulosa Torr. (honey mesquite)

Prosopis sp. (mesquite)

Quercus fusiformis [Small] Sarg. (scrub live oak, plateau oak)

Quercus mohriana Buckl. (mohrs shin oak)

Quercus sp. (oak)

Setaria leucopila [Schribn. & Merr.] K. Schumann (bristle grass, millet)

Setaria lutescens [Weigel] Hubb. (yellow foxtail, pigeon grass, yellow bristle grass)

Sophora secundiflora [Ort.] Lag ex DC. (mescal bean, Texas mountain laurel, frijolito)

Sporobolus cryptandrus [Torr.] Gray (covered-spike dropseed, sand dropseed)

Sporobolus sp. (dropseed)

Tripsacum dactyloides [L.] L. (eastern gamagrass)

Ungnadia speciosa Endl. (Mexican buckeye)

Xanthium pensylvanicum Wallar. (cocklebur)

Xanthium sp. (cocklebur)

Yucca sp. (yucca)

I. FATE BELL SHELTER (41 VV 74)

SEEDS, PODS, NUTS, AND FRUIT

Approximate Descending Order of Relative Frequency of Occurrence:

Quercus mohriana	(6)
Juglans microcarpa	(6)
Sophora secundiflora	(3)
Quercus sp.	(1)
Opuntia sp.	(1)
Ungnadia speciosa	(1)
Celtis reticulata	(1)

Intrasite Distribution:

Random and miscellaneous:

Juglans microcarpa	(2)

Zone I:

Quercus mohriana	(5)
Juglans microcarpa	(1)
Sophora secundiflora	(1)

Zone III, Level I:

Quercus mohriana	(1)

Zone III, Level II:

Juglans microcarpa	(1)
Quercus sp.	(1)
Sophora secundiflora	(1)

Zone IV:

Juglans microcarpa	(2)
Opuntia sp.	(1)
Sophora secundiflora	(1)
Ungnadia speciosa	(1)
Celtis reticulata	(1)

LEAVES, FIBERS, AND STEMS

Approximate Descending Order of Relative Frequency of Occurrence:

Agave lecheguilla	(10)
Yucca sp.	(3)
Opuntia sp.	(3)
Sophora secundiflora	(2)
Ariocarpus sp.	(1)

Intrasite Distribution:

Random and miscellaneous:

Ariocarpus sp.	(1)
Yucca sp.	(1)

Zone I:

Agave lechequilla	(3)
Opuntia sp.	(2)

Zone III, Level I:

Agave lechequilla	(2)
Yucca sp.	(1)
Sophora secundiflora	(1)

Zone III, Level II:

Agave lecheguilla   (1)

Yucca sp.   (1)

Zone III, general:

Agave lecheguilla   (1)

Zone IV:

Agave lecheguilla   (2)

Mixed Zones III and IV:

Agave lecheguilla   (1)

Opuntia sp.   (1)

Sophora secundiflora   (1)

II. COONTAIL SPIN SITE (41 VV 82)

SEEDS, PODS, NUTS, AND FRUIT

Approximate Descending Order of Relative Frequency of Occurrence:

Prosopis glandulosa	(27)
Juglans microcarpa	(20)
Quercus mohriana	(18)
Acacia berlandieri	(9)
Leucaena retusa	(8)
Yucca sp.	(8)
Ungnadia speciosa	(7)
Sophora secundiflora	(6)
Diospyros texana	(5)
Celtis reticulata	(5)
Quercus sp.	(4)
Acacia rigidula	(3)
Helianthus sp.	(3)
Cassia roemeriana	(2)
Agave lecheguilla	(2)
Quercus fusiformis	(2)
Opuntia sp.	(2)
Juglans sp.	(2)
Acacia greggii	(1)
Karwinskia humboldtiana	(1)
Xanthium sp.	(1)

Intrasite Distribution:

Surface:

Prosopis glandulosa	(5)
Sophora secundiflora	(4)
Leucaena retusa	(1)
Yucca sp.	(1)
Quercus sp.	(1)
Acacia berlandieri	(1)

Random and miscellaneous:

Quercus mohriana	(4)
Prosopis glandulosa	(3)
Juglans microcarpa	(3)
Leucaena retusa	(2)
Acacia rigidula	(2)
Acacia berlandieri	(2)
Ungnadia speciosa	(1)
Acacia greggii	(1)
Quercus fusiformis	(1)
Cassia roemeriana	(1)
Sophora secundiflora	(1)

Area A:

Upper A-3:

Prosopis grandulosa	(2)
Juglans microcarpa	(2)
Celtis reticulata	(2)
Quercus mohriana	(1)

Upper A-3, continued

<u>Diospyros</u> <u>texana</u>	(1)
<u>Acacia</u> <u>berlandieri</u>	(1)
<u>Helianthus</u> sp.	(1)
<u>Opuntia</u> sp.	(1)
<u>Ungnadia</u> <u>speciosa</u>	(1)
<u>Karwinskia</u> <u>humboldtiana</u>	(1)
<u>Leucaena</u> <u>retusa</u>	(1)
<u>Agave</u> <u>lecheguilla</u>	(1)

Lower A-3:

<u>Quercus</u> <u>mohriana</u>	(2)
<u>Yucca</u> sp.	(1)
<u>Prosopis</u> <u>grandulosa</u>	(1)
<u>Leucaena</u> <u>retusa</u>	(1)

Transitional:

<u>Prosopis</u> <u>grandulosa</u>	(6)
<u>Juglans</u> <u>microcarpa</u>	(4)
<u>Yucca</u> sp.	(3)
<u>Diospyros</u> <u>texana</u>	(2)
<u>Acacia</u> <u>berlandieri</u>	(2)
<u>Quercus</u> <u>mohriana</u>	(2)
<u>Celtis</u> <u>reticulata</u>	(2)
<u>Leucaena</u> <u>retusa</u>	(1)
<u>Quercus</u> <u>fusiformis</u>	(1)
<u>Quercus</u> sp.	(1)
<u>Ungnadia</u> <u>speciosa</u>	(1)
<u>Juglans</u> sp.	(1)
<u>Opuntia</u> sp.	(1)

Upper A-4:

Prosopis grandulosa	(4)
Quercus mohriana	(3)
Ungnadia speciosa	(2)
Helianthus sp.	(1)
Acacia berlandieri	(1)
Acacia rigidula	(1)
Diospyros texana	(1)
Juglans microcarpa	(1)
Yucca sp.	(1)
Leucaena retusa	(1)

Middle A-4:

Quercus mohriana	(2)
Juglans microcarpa	(2)
Prosopis grandulosa	(1)

Lower A-4:

Juglans microcarpa	(3)
Sophora secundiflora	(2)
Acacia berlandieri	(2)
Quercus mohriana	(2)
Prosopis grandulosa	(2)
Helianthus sp.	(1)
Ungnadia speciosa	(1)
Quercus sp.	(1)
Leucaena retusa	(1)
Agave lecheguilla	(1)

6'-7' below datum:

Juglans microcarpa	(2)
Xanthium sp.	(1)
Ungnadia speciosa	(1)
Diospyros texana	(1)
Celtis reticulata	(1)
Prosopis grandulosa	(1)
Quercus sp.	(1)

8'-9' below datum:

Juglans microcarpa	(1)
Prosopis grandulosa	(1)

11'-12' below datum:

Leucaena retusa	(1)
Quercus mohriana	(1)
Prosopis grandulosa	(1)
Cassia roemeriana	(1)

Area B:

2'-3' below surface:

Sophora secundiflora	(2)

3'-4' below surface:

Juglans microcarpa	(1)
Quercus mohriana	(1)

4'-5' below surface:

Yucca sp.	(2)
Juglans microcarpa	(1)
Juglans sp.	(1)

BULBS

Allium drummondi   (5)

Intrasite Distribution:

  Surface:

    Allium drummondi   (1)

  Random and miscellaneous:

    Allium drummondi   (1)

  Area A:

    Upper A-3:

      Allium drummondi   (1)

    Transitional:

      Allium drummondi   (2)

QUIDS? (Masses of Fiber, Possibly Chewed)

Approximate Descending Order of Relative Frequency of
Occurrence:

        Agave lecheguilla   (65)

        Sporobolus sp.   (3)

Intrasite Distribution:

  Surface:

        Agave lecheguilla   (9)

  Random and miscellaneous:

        Agave lecheguilla   (13)

  Area A:

    Upper A-3:

        Agave lecheguilla   (2)

        Sporobolus sp.   (1)

Lower A-3:

    Agave *lecheguilla*   (2)

Transitional:

    Agave *lecheguilla*   (9)

    Sporobolus sp.   (1)

Upper A-4:

    Agave *lecheguilla*   (8)

Middle A-4:

    Agave *lecheguilla*   (2)

Lower A-4:

    Agave *lecheguilla*   (3)

6'-7' below datum:

    Agave *lecheguilla*   (3)

7'-8' below datum:

    Agave *lecheguilla*   (1)

Area B:

1'-2' below surface:

    Agave *lecheguilla*   (1)

2'-3' below surface:

    Agave *lecheguilla*   (7)

3'-4' below surface:

    Agave *lecheguilla*   (4)

    Sporobolus sp.   (1)

5'-6' below surface:

    Agave *lecheguilla*   (1)

LEAVES, STEMS, CULMS, AND FLOWERS

Approximate Descending Order of Relative Frequency of
Occurrence:

Agave lecheguilla	(14)
Setaria leucopila	(11)
Sporobolus sp.	(10)
Sophora secundiflora	(7)
Opuntia lindheimeri	(3)
Yucca sp.	(2)
Opuntia sp.	(2)
Soporbolus cryptandrus	(2)
Setaria sp.	(2)
Pappophorum bicolor	(1)
Celtis sp.	(1)
Setaria lutescens	(1)
Quercus fusiformis	(1)
Diospyros texana	(1)
Tripsacum dactyloides	(1)
Aristida glauca	(1)

Intrasite Distribution:

   Surface:

Agave lecheguilla	(1)
Sporobolus sp.	(1)

   Random and miscellaneous:

Sporobolus sp.	(3)

Setaria leucopila	(3)
Agave lecheguilla	(3)
Pappophorum bicolor	(1)
Sporobolus cryptandrus(?)	(1)
Yucca sp.	(1)
Celtis sp.	(1)
Setaria lutescens	(1)

Area A:

Lower A-3:

Sophora secundiflora	(2)
Quercus fusiformis	(1)
Diospyros texana	(1)

Transitional:

Sporobolus sp.	(2)
Agave lecheguilla	(2)
Setaria leucopila	(2)
Opuntia lindheimeri	(1)
Tripsacum dactyloides	(1)
Aristida glauca	(1)

Upper A-4:

Setaria leucopila	(3)
Sophora secundiflora	(1)
Yucca sp.	(1)
Agave lecheguilla	(1)
Sporobolus sp.	(1)

Middle A-4:

> Agave lecheguilla   (1)
>
> Sporobolus sp.   (1)
>
> Setaria leucopila   (1)
>
> Opuntia lindheimeri   (1)

Lower A-4:

> Sophora secundiflora   (2)
>
> Agave lecheguilla   (2)
>
> Sporobolus cryptandrus   (1)
>
> Sporobolus sp.   (1)
>
> Setaria leucopila   (1)
>
> Opuntia lindheimeri   (1)

6'-7' below datum:

> Agave lecheguilla   (2)
>
> Setaria sp.   (1)

8'-9' below datum:

> Opuntia sp.   (1)

Area B:

2'-3' below surface:

> Setaria sp.   (1)
>
> Agave lecheguilla   (1)
>
> Sporobolus sp.   (1)
>
> Opuntia sp.   (1)

3'-4' below surface:

    Agave lecheguilla   (1)

    Sophora secundiflora   (1)

5'-6' below surface:

    Setaria sp.   (1)

III. ZOPILOTE CAVE (41 VV 216)

SEEDS, PODS, NUTS, AND FRUIT

Approximate Descending Order of Relative Frequency of Occurrence:

    Juglans microcarpa   (12)

    Prosopis grandulosa   (7)

    Quercus fusiformis   (5)

    Acacia berlandieri   (4)

    Sophora secundiflora   (3)

    Leucaena retusa   (3)

    Quercus sp.   (2)

    Ungnadia speciosa   (2)

    Diospyros texana   (1)

    Quercus mohriana   (1)

Intrasite Distribution - not listed because deposit undifferentiated

POSSIBLE QUIDS

Agave lecheguilla   (18)

Intrasite Distribution - not listed because deposit undifferentiated.

LEAVES, STEMS, CULMS, AND FLOWERS

Approximate Descending Order of Relative Frequency of Occurrence:

Agave lecheguilla	(22)
Opuntia sp.	(19)
Opuntia lindheimeri	(15)
Opuntia phaeacantha	(12)
Sophora secundiflora	(3)
Agave sp.	(2)
Sporobolus sp.	(1)
Yucca sp.	(1)

Intrasite Distribution - not listed because deposit undifferentiated.

BULB

Allium drummondi	(1)

IV.   EAGLE CAVE (41 VV 167)

SEEDS, PODS, NUTS, AND FRUIT

Quercus fusiformis	(65)
Juglans microcarpa	(61)
Ungnadia speciosa	(13)
Prosopis glandulosa	(13)

Opuntia sp.   (12)

Yucca sp.   (9)

Sophora secundiflora   (7)

Leucaena retusa   (4)

Diospyros texana   (3)

Acacia rigidula   (3)

Quercus sp.   (3)

Celtis reticulata   (2)

Acacia roemeriana   (1)

Cucurbita foetidissima   (1)

Prosopis sp.   (1)

Opuntia lindheimeri   (1)

Intrasite Distribution:

Random and miscellaneous:

Quercus fusiformis   (9)

Juglans microcarpa   (4)

Ungnadia speciosa   (3)

Opuntia sp.   (2)

Prosopis grandulosa   (1)

Celtis reticulata   (1)

Acacia rigidula   (1)

Stratum I:

Sophora secundiflora   (3)

Prosopis grandulosa   (3)

Juglans microcarpa   (3)

Yucca sp.	(2)
Opuntia sp.	(2)
Leucaena retusa	(1)
Quercus fusiformis	(1)
Ungnadia speciosa	(1)
Opuntia lindheimeri	(1)

Stratum IIa:

Prosopis grandulosa	(6)
Yucca sp.	(5)
Quercus fusiformis	(4)
Juglans microcarpa	(3)
Opuntia sp.	(3)
Sophora secundiflora	(3)
Quercus sp.	(2)
Diospyros texana	(2)
Celtis reticulata	(1)
Cucurbita foetidissima	(1)
Leucaena retusa	(1)
Acacia roemeriana	(1)
Ungnadia speciosa	(1)
Acacia rigidula	(1)
Prosopis sp.	(1)

Strata IIc and IId:

Quercus fusiformis	(24)
Juglans microcarpa	(19)
Ungnadia speciosa	(3)

       Opuntia sp.   (2)

       Quercus sp.   (1)

       Prosopis grandulosa   (1)

       Acacia rigidula   (1)

       Yucca sp.   (1)

       Sophora secundiflora (1)

       Leucaena retusa   (1)

Stratum II - general:

       Juglans microcarpa   (2)

Mixed Strata II and III:

       Juglans microcarpa   (1)

Stratum III:

       Juglans microcarpa   (19)

       Quercus fusiformis   (19)

       Ungnadia speciosa   (5)

       Opuntia sp.   (2)

       Yucca sp.   (1)

       Diospyros texana   (1)

       Prosopis grandulosa   (1)

Stratum IV:

       Quercus fusiformis   (4)

       Juglans microcarpa   (3)

Mixed Strata III and IV:

       Juglans microcarpa   (1)

Stratum V:

Juglans microcarpa	(6)
Quercus fusiformis	(4)
Prosopis grandulosa	(1)
Opuntia sp.	(1)
Leucaena retusa	(1)

LEAVES, STEMS, AND FIBERS

Approximate Descending Order of Relative Frequency of Occurrence:

Agave lecheguilla	(47)
Yucca sp.	(4)
Opuntia sp.	(3)
Quercus fusiformis	(2)
Opuntia lindheimeri	(2)
Ariocarpus fissuratus	(1)
Karwinskia humboldtiana (?)	(1)
Opuntia phaeacantha	(1)
Sophora secundiflora	(1)

Intrasite Distribution:

Random and miscellaneous:

Agave lecheguilla	(5)
Opuntia lindheimeri	(1)

Stratum I:

Agave lecheguilla	(4)
Yucca sp.	(3)

Ariocarpus fissuratus   (1)

Opuntia lindheimeri   (1)

Stratum IIa:

Agave lecheguilla   (7)

Opuntia phaecantha   (1)

Sophora secundiflora   (1)

Quercus fusiformis   (1)

Karwinskia humboldtiana(?)(1)

Stratum IIb:

Agave lecheguilla   (2)

Quercus fusiformis   (1)

Strata II and IId:

Agave lecheguilla   (10)

Opuntia sp.   (2)

Yucca sp.   (1)

Stratum III:

Agave lecheguilla   (12)

Stratum IV:

Agave lecheguilla   (1)

Stratum V:

Agave lecheguilla   (6)

Opuntia sp.   (1)

BULB

Allium drummondi   (1)

Intrasite Distribution:

    Random and miscellaneous:

        <u>Allium</u> <u>drummondi</u>   (1)

QUIDS (?)

Approximate Descending Order of Relative Frequency of
Occurrence:

        <u>Agave</u> <u>lecheguilla</u>   (7)

Intrasite Distribution:

    Stratum IId:

        <u>Agave</u> <u>lecheguilla</u>   (2)

    Mixed Strata II and III:

        <u>Agave</u> <u>lecheguilla</u>   (1)

    Stratum III:

        <u>Agave</u> <u>lecheguilla</u>   (3)

    Stratum IV:

        <u>Agave</u> <u>lecheguilla</u>   (1)

V.   DEVIL'S MOUTH SITE (41 VV 188)

SEEDS AND SEED PODS

    <u>Jatropha</u> <u>dioica</u>   (3)

Intrasite Distribution:

    Stratum 1:

        <u>Jatropha</u> <u>dioica</u>   (1)

    Stratum 3:

        <u>Jatropha</u> <u>dioica</u>   (2)

VI. BONFIRE SHELTER (41 VV 218)

SEEDS, PODS, NUTS, AND FRUIT

Approximate Descending Order of Relative Frequency of Occurrence:

Yucca sp.	(16)
Juglans microcarpa	(11)
Celtis reticulata	(8)
Prosopis grandulosa	(7)
Opuntia sp.	(7)
Quercus fusiformis	(5)
Leucaena retusa	(3)
Xanthium pensylvanicum	(2)
Sophora secundiflora	(2)
Acacia rigidula	(2)
Agave sp.	(1)
Carya sp.	(1)
Cucurbita foetidissima	(1)
Prosopis sp.	(1)

Intrasite Distribution:

Random and miscellaneous:

Juglans microcarpa	(1)

Fiber Layer:

Yucca sp.	(15)
Juglans microcarpa	(8)

Opuntia sp.	(7)
Prosopis grandulosa	(7)
Quercus fusiformis	(5)
Leucaena retusa	(3)
Acacia rigidula	(2)
Celtis reticulata	(1)
Sophora secundiflora	(1)
Agave sp.	(1)
Prosopis sp.	(1)
Carya sp.	(1)
Cucurbita foetidissima	(1)

Possibly from Fiber Layer:

Yucca sp.	(1)

Mixed Zone III and Bone Bed 3:

Juglans microcarpa	(1)
Xanthium pensylvanicum	(1)

Bone Bed 3:

Celtis reticulata	(1)
Juglans microcarpa	(1)
Xanthium pensylvanicum	(1)

Bone Bed 2:

Celtis reticulata	(5)
Sophora secundiflora	(1)

Zone I:

Celtis reticulata	(1)

LEAVES, STEMS, CULMS, AND FLOWERS

Approximate Descending Order of Relative Frequency of
Occurrence:

Agave lecheguilla	(10)
Sophora secundiflora	(5)
Tripsacum dactyloides	(4)
Sporobolus sp.	(3)
Celtis reticulata	(2)
Yucca sp.	(2)
Leucaena retusa	(1)
Diospyros texana	(1)

Intrasite Distribution:

Fiber Layer:

Agave lecheguilla	(10)
Tripsacum dactyloides	(4)
Sporobolus sp.	(3)
Yucca sp.	(2)
Celtis reticulata	(1)
Sophora secundiflora	(1)

Possibly from Fiber Layer:

Sophora secundiflora	(2)
Leucaena retusa	(1)

Zone III - general:

Sophora secundiflora	(1)

Bone Bed 3:

Diospyros texana	(1)
Sophora secundiflora	(1)
Celtis reticulata	(1)

DISCUSSION OF MACROFOSSILS

Interpretations

The preliminary reports indicate that plant usage, as recorded in the fossil record, differed among the six sites some of which are separated by as much as 40 miles. An example of this is a comparison between Eagle Cave and Coontail Spin during time Periods II-III. At Eagle Cave the predominant food record is one of walnut and acorns of Quercus fusiformis while Coontail Spin, 40 miles distant, has fewer walnuts and acorns of Quercus mohriana.

Cactus seeds occur randomly throughout the six sites, but in Zopilote Cave, a concentration of cactus pads occurs throughout the strata.

While primitive man used Agave lecheguilla extensively, leaf fragments and quids are most abundant in Zopilote, Coontail Spin, and Eagle Cave. Discrimination against sotol (Dasylirion), Yucca, and ocotillo is evidences by either a total absence or rarity of occurrence in relation to Agave and others. It has been suggested that Agave lecheguilla was being used for fermentation, curing of hides, or as a direct source of food. Further work is needed to determine its actual use.

The occurrence of Sophora seed and seed coat fragments in all shelter sites (Eagle, Zopilote, Coontail Spin, Bonfire, and Fate Bell) coupled with ethnographic data suggests the possibility that primitive men living in the Rio Grande Trans-Pecos region were using the seed coats for their hallucenogenic properties.

Cautious interpretation and much additional work is needed because of some obvious rodent activity. However, human teeth impressions in Agave quids give rather strong evidence that human factors are important in determining the plant macrofossil record. Additional supporting data should be sought by careful and understanding archeological excavation techniques. Additionally, the habits of contemporary indigenous animals needs investigation to provide data which can be used to exclude or confirm animal disturbance and emplanting of plant remains.

Suggestions on Future Sampling

There are some suggestions which might be made from the identifying end of the project on sampling.

The first of such would be that more of the sample from a given site be available for checking by the identifier. In many cases in the past work, identification had to be made from 1 or 2 seeds or 1 leaflet. A larger sample might have given more of the specific sample and more certainty in the identification. This is especially true where heterogeneity occurs in the sample, i.e., many taxa but thinly represented.

Another suggestion might be that whenever possible the screening be done by the identifier. Admittedly, this would entail more work, but such a procedure would perhaps lead to the uncovering of new taxa or at least the assembling of a more complete list for any given site.

Finally, it might be noted that a uniform numbering system of the samples should be attained. In much of the above work, lot numbers were used for the presentation of identifications, but in many others, such information was missing. Perhaps a separate identification number could be employed.

THE AMISTAD POLLEN REFERENCE COLLECTION

Vaughn M. Bryant, Jr.

Preparations of the Amistad Pollen Reference Collection were initiated during the summer of 1965 by Drs. Donald A. Larson, John H. McAndrews, and the author, and have continued through to the present. Approximately 350 of the samples in the collection are of pollen extracted from specimens on herbarium sheets in The University of Texas Herbarium (see list attached). Most of the pollen types selected for inclusion in the reference collection are from: 1) plant types known to exist in the Amistad region, 2) plant taxa growing in regions bordering the Amistad Reservoir, and 3) plants such as _Picea_, _Pseudotsuga_, _Pinus_, and _Ephedra nevadensis_ which do not now live in or near the Amistad region but appear in the fossil pollen record.

The herbarium specimens were processed in The University of Texas Palynological Laboratory by John McAndrews, Ann Walters, and the author, using standard techniques outlined by Faegri and Iversen (1964). After processing, the samples were placed in silicone oil, mounted on glass slides and labeled (see Appendix B of the Devil's Mouth Site report, herein, for details on silicone oil mounting techniques).

The Amistad Reference Collection contains approximately 250 additional pollen slides which were obtained through exchanges with other palynological laboratories. These additional slides are currently being analyzed and cataloged and will soon be incorporated into a preliminary key of Southwest Texas pollen types.

POLLEN TYPES FROM
THE UNIVERSITY OF TEXAS HERBARIUM*

ACANTHACEAE (Acanthus Family)

 Anisacanthus insignis 279** (dwarf anisacanth)

 Ruellia yucatana 267

ACERACEAE (Maple Family)

 Acer grandidentatum 256 (bigtooth maple)

AMARANTHACEAE (Amaranthus Family)

 Amaranthus berlandieri 5 (berlandier amaranthus)

 Amaranthus palmeri 157 (careless weed, palmer
      amaranthus)

 Amaranthus retroflexus 4 (pigweed, red root,
      green amaranthus)

 Amaranthus tamariscinus 54 (water hemp)

 Brayulinea densa 56

 Celosia nitida 46 (cockscomb)

 Froelichia campestris 98 (Florida snake cotton)

 Froelichia drummondii 94, 96 (drummond snake cotton)

 Froelichia gracilis 90, 92 (slender snake cotton)

 Gossypianthus lanuginosus 36 (woolly cotton flower)

 Iresine celosia 34 (jubas bush)

 Lagrezia monosperma 22

 Nototrichium sandwicense 16

---

 *Common names and spelling after Schulz (1928), Little
(1950), and Gould (1962).

 **University of Texas Palynological Laboratory Accession
Number.

Philoxerus vermicularis 24 (silverhead)

Tidestromia lanuginosa 28 (woolly tidestromia)

AMARYLLIDACEAE (Amaryllis Family)

Agave lecheguilla 15 (lechuguilla)

Hymenocallis occidentalis 33 (spider lily, inland
hymencallis)

ANACARDIACEAE (Cashew Family)

Rhus trilobata 200 (three-leaved sumac, skunk bush,
squash bush)

ARISTOLOCHIACEAE (Birthwort Family)

Aristolochia coryi 63 (cory dutchman's pipe)

BERBERIDACEAE (Barberry Family)

Berberis trifoliolata 55 (agarita, wild currant)

BETULACEAE (Birch Family)

Alnus crispa 320 (mountain alder)

Alnus rugosa 296 (speckled alder)

Betula glandulosa 312 (tundra dwarf birch)

BIGNONIACEAE (Bignonia Family)

Chilopsis linearis 115 (desert willow, flowering
willow)

BORAGINACEAE (Borage Family)

Borrago officinalis 121

Coldenia hispidissima 219 (rough coldenia)

Heliotropium angustifolium 222 (slim-leaf heliotrope)

Lappula redowskii 213 (flat-spine stickseed)

Lappula texana 223 (bur forget-me-not, hairy stickseed)

Lithospermum multiflorum 212 (many-flower gromwell)

Tournefortia hartwegiara 215

CACTACEAE (Cactus Family)

Echinocactus texensis 204 (devil's head, devil's
pincushion)

Echinocereus blanckii 177 (blanck's echinocereus)

Mammillaria heyderi 184 (devil's pincushion, heyder
mammillaria)

Neomammillaria hebrichiara 194

Opuntia arbuscula 198

Opuntia leptocaulis 188, 207 (tasajillo, rattail
cactus, Christmas cactus)

Opuntia lindheimeri 196 (nopal prickly pear, Texas
prickly pear)

Rhipsalis cassytha 187

CAMPANULACEAE (Bellflower Family)

Lobelia brachypoda 286

CAPPARIDACEAE (Caper Family)

Cleome gynandra 9 (prickly spiderflower)

CAPRIFOLIACEAE (Honeysuckle Family)

Lonicera albiflora 276 (white honeysuckle)

Sambucus coerulea 285 (Mexican elder, elderberry,
blueberry elder, tapiro)

Symphoricarpus orbiculatus 271 (coralberry, Indian
currant, buckbush)

CARYOPHYLLACEAE (Pink Family)

    Drymaria debilis 234

CELASTRACEAE (Staff-tree Family)

    Schaefferia cuneifolia 252 (redberry, desert yaupon)

CHENOPODIACEAE (Goosefoot Family)

    Allenrolfea occidentalis 10 (pickleweed)

    Atriplex acanthocarpa 12 (armed saltbush)

    Atriplex canescens 18 (four-wing saltbush, chamiza)

    Atriplex pentandra 30

    Atriplex texana 38 (Texas saltbush)

    Chenopodium album 137 (white goosefoot, wild spinach, pigweed, frostbit, bacon weed, lamb's quarters)

    Chenopodium ambrosiodes 48, 150 (wormseed goosefoot, wormseed lamb's quarters, Mexican tea, Spanish tea)

    Chenopodium atrovirens 152

    Chenopodium berlandieri 52 (pigseed goosefoot)

    Chenopodium botrys 127 (feather geranium)

    Chenopodium bushianum 135

    Chenopodium californicum 112

    Chenopodium capitatum 124

    Chenopodium cycloides 151

    Chenopodium desiccatum 64 (thick-leaf goosefoot)

    Chenopodium fremontii 140 (fremont goosefoot)

Chenopodium glaucum 118

Chenopodium gigantospermum 72, 159 (bigseed goosefoot)

Chenopodium graveolens 148 (scented goosefoot)

Chenopodium hians 141

Chenopodium hybridum 160

Chenopodium incanum 142 (mealy goosefoot)

Chenopodium leptophyllum 7 (slim-leaf goosefoot, narrow-leaf goosefoot, narrow-leaf lamb's quarters)

Chenopodium macrospermum 128

Chenopodium missouriense 143 (Missouri goosefoot)

Chenopodium murale 78 (nettel-leaf goosefoot)

Chenopodium neomexicanum 144

Chenopodium nevadense 136

Chenopodium pahuense 146

Chenopodium opulifolium 129

Chenopodium overi 126

Chenopodium polyspermum 130

Chenopodium pumilio 119

Chenopodium rubrum 2, 158

Chenopodium serotinum 138

Chenopodium standleyanum 131 (standley goosefoot)

Chenopodium strictum 132

Chenopodium vulvaria 133

Cycloloma atriplicifolium 155 (plains tumbleweed, winged pigweed, tumble ringwing)

Eurotia ceratoides 50

Eurotia lanata 153 (common winterfat, white sage)

Grayia spinosa 74

Gomphrena decumbens 40

Halocnemon strobilaceum 62

Halogeton glomeratus 26

Haloxylon articulatum 68

Helmbergia twedei 42

Kochia americana 76

Kochia scoparia 154 (belvedere summer cypress)

Salicornia perennis 84 (woody glasswort, bush glasswort)

Salsola kali 86 (Russian thistle, tumbleweed)

Sarobatus vermiculatus 308

Schanginia baccata 88

Spirostachys sp. 66

Suadea conferta 60

COMPOSITAE (Composite Family)

Ambrosia artemisiifolia 298 (common ragweed, Roman
wormwood)

Ambrosia confertiflora 299 (field ragweed, bursage)

Artemisia dracunculoides 310 (false tarragon, sagewort)

Artemisia filifolia 340 (sand sagebrush)

Artemisia mexicana 305 (Mexican sagewort, wormwood)

Artemisia tridentata 321 (sagebrush, wormwood)

Crepis petiolata 280 (hawksbeard)

Flourensia cernau 147 (tarbush, varnishbush, blackbush)

Franceria acanthicarpa 334 (burweed)

Guterezia texana 337 (broomweed, kindlingweed)

Helianthus mollis 342 (ashy sunflower, hairy sunflower)

Hymenoclea salsola 335 (burrobush, cheeseweed)

Iva ambrosiaefolia 180 (rag sumpweed)

Iva frutescens 301 (big-leaf sumpweed, marshelder)

Iva texensis 170 (Texas sumpweed)

Lindheimera texana 300 (Texas star, Lindheimer's daisy,
                          Texas star daisy, yellow Texas
                          star)

Lygodesmia texana 336 (Texas skeleton plant)

Sonchus oleraceus 343 (common sow thistle)

Xanthium pensylvanicum 341 (cocklebur)

Zinnia acerosa 297 (spiney-leaf zinnia)

CONVOLVULACEAE (Morning-Glory Family)

Dichondra brachypoda 191 (New Mexico pony-foot)

Ipomoea costellata 167 (crestrib morning-glory)

CORNACEAE (Dogwood Family)

Garrya lindheimeri 293

CRUCIFERAE (Mustard Family)

Halimolobos diffusus 311

CUCURBITACEAE (Gourd Family)

Cucurbita foetidissima 183 (buffalo gourd, stinking
                            gourd, mock orange, cala-
                            bazilla, Missouri gourd,
                            fetid wild pumpkin)

Cucurbita texana 181

Ibervillea tenuisecta 171 (slim-lobe globe berry,
                          deer apple)

Sicyos parviflorus 162 (small flower bur-cucumber)

CYPERACEAE (Sedge Family)

Carex pennsylvanica 25

Cyperus laevigatus 326 (smooth flat-sedge)

DROSERACEAE (Sundew Family)

Drosera annua 309 (annual sundew)

EBENACEAE (Elbony Family)

Diospyros texana 262 (Texas persimmon, Mexican per-
                          simmon, black persimmon)

ERICACEAE (Heath Family)

Arbutus menziesii 290

Arbutus texana 294 (Texas madrona, naked Indian,
                        Texas arbutus)

EPHEDRACEAE/GNETACEAE (Ephedra Family/Joint-fir Family)

Ephedra antisyphilitica 149 (vine joint-fir, vine
                              ephedra)

Ephedra aspera 241, 330 (boundary ephedra)

Ephedra californica 331 (California ephedra)

Ephedra nevadensis 315 (Nevada ephedra, Nevada joint-fir)

Ephedra torreyana 319 (Morman tea, torrey ephedra, torrey
                     joint-fir)

Ephedra trifurca 240 (Mexican tea, long-leaf ephedra)

EUPHORBIACEAE (Spurge Family)

Euphorbia mauritanica 244

Jatropha dioica 243 (leather stem, sangre de drago,
rubber plant)

Phyllanthus polygonoides 242 (knotweed leaf-flower)

FAGACEAE (Beech Family)

Quercus emoryi 169 (emory oak, emory's black oak,
blackjack oak)

Quercus grisea 161 (gray oak, mountain white oak)

Quercus marilandica 201 (blackjack oak)

Quercus virginiana 185 (live oak)

FOUQUIERIACEAE (Coach-Whip Family)

Fouquieria splendens 51 (candlewood, Jacob's staff,
ocitillo, coach-whip cactus,
slim-wood)

GENTIANACEAE (Gentian Family)

Eustoma exaltatum 295 (tall prairiegentian, catchfly
gentian)

Sabatia campestris 120 (sea star, meadow pink, rose
pink, prairie rosegentian)

GERANIACEAE (Geranium Family)

Erodium texanum 116 (stork's bill, heron bill, Texas
filaree, wild geranium, pine
needle)

GRAMINEAE (Grass Family)

Andropogon sp. 325 (bluestem)

Stenotaphrum secundatum 323 (Saint Augustine grass, shoreline)

Tripsacum dactyloides 322 (eastern gama grass)

HAMAMELIDACEAE (Witch-hazel Family)

Liquidambar styraciflua 307 (sweet gum)

HYDROPHYLLACEAE (Waterleaf Family)

Nama hispidum 291 (sand bells, rough nama)

Phacelia congesta 113 (blue curls, spider flower, caterpillars, wild helio-trope, snail flower)

JUGLANDACEAE (Walnut Family)

Carya texana 205 (black hickory, Buckley's hickory)

Juglans microcarpa 61 (little walnut, Texas black walnut)

KOEBERLINIACEAE (Junco Family or Allthorn Family)

Koeberlinia spinosa 45 (crown of thorns, junco, allthorn, crucifixion thorn)

LABIATAE (Mint Family)

Marrubium vulgare 272 (common horehound)

Salazaria mexicana 280 (bladder sage)

LEGUMINOSAE (Pea Family)

Acacia greggii 174 (cat's claw, devil's claw, chaparral)

Acacia texensis 165 (acacia)

Brogniartia sp. 220

Calliandra eriophylla 228

Cassia bauhinioides 229 (shubby senna)

Cercis canadensis 250 (judas tree, eastern redbud)

Cercis occidentalis 247 (judas tree, Californian
redbud, Arizona redbud)

Dalea frutescens 208 (shrubby dalea, black dalea)

Leucaena retusa 210 (little-leaf leadtree)

Krameria glandulosa 249

Mimosa biuncifera 218 (mimosa, cat's claw mimosa,
wait-a-minute, wait-a-bit)

Prosopis glandulosa 95 (honey mesquite)

Sophora secundiflora 216 (Texas mountain laurel, big
drunk bean, mescal bean,
frijolito)

LILIACEAE (Lily Family)

Dasylirion leiophyllum 57 (smooth sotol, sow yucca,
spoonplant)

Nolina texana 49 (slender bear grass, basket grass,
sacahuiste)

Yucca arkansana 11 (Arkansas yucca)

Yucca baccata 289 (banana yucca, blue yucca, datil
yucca)

LOASACEAE (Loasa Family)

Mentzelia multiflora 43 (desert mentzelia)

LOGANIACEAE (Logania Family)

Buddleia marrubiifolia 292 (woolly butterfly bush)

LORANTHACEAE (Mistletoe Family)

    Phoradendron _coryae_ 175 (oak mistletoe)

LYTHRACEAE (Loosestrife Family)

    Heimia _longipes_ 261 (stalkflower heimia)

MALPIGHIACEAE (Malpighia Family)

    Janusia _gracilis_ 251 (slender janusia)

MALVACEAE (Mallow Family)

    Abutilon _incanum_ 47, 173 (Indian mallow)

    Gossypium _hirsutum_ 17

    Hibiscus _coulteri_ 189 (desert rose mallow)

    Sida _neomexicana_ 288 (New Mexico sida)

    Sphaeralcea _angustifolia_ 39, 327 (narrow-leaf globe
                               mallow)

    Sphaeralcea _coccinea_ 328 (scarlet globe mallow, red
                               false mallow)

    Sphaeralcea _hastulata_ 329 (spear globe mallow)

MENISPERMACEAE (Moonseed Family)

    Cocculus _carolinus_ 230 (coral-bead, wild sarsaparilla,
                            redberried moonseed, margil,
                            coral-vine)

MORACEAE (Mulberry Family)

    Maclura _pomifera_ 239 (osage orange, horse apple,
                      bois d'arc)

    Morus _microphylla_ 59 (Texas mulberry, mountain mulberry)

104

NYCTAGINACEAE (Four-o'clock Family)

    Abronia angustifolia 221 (narrow-leaf sand verbena)

    Allionia incarnata 224 (trailing four-o'clock, trailing allionia)

    Anulocaulis leisolenus 226 (ring stem)

    Boerhaavia erecta 23 (erect spiderling)

    Cyphomeris sp. 217

    Nyctaginia capitata 344 (devil's bouquet, scarlet muckflower)

    Oxybaphus albidus 29

    Pisonia aculeata 345

    Selinocarpus angustifolius 214 (narrow-leaf moonpod)

ONAGRACEAE (Evening Primrose Family)

    Gaura coccinea 186 (scarlet gaura)

    Guara macrocarpa 202 (transpecos guara)

    Jussiaea diffusa 163 (floating water primrose, water primrose)

    Oenothera albicaulis 168 (pale evening primrose)

    Oenothera lampasana 182 (grand prairie evening primrose)

    Oenothera speciosa 172 (showy sundrops, showy primrose, white evening primrose)

OLEACEAE (Olive Family)

    Menodora heterophylla 106 (redbud, menodora)

    Forestiera angustifolia 85

    Fraxinus texensis 65 (ash)

    Fraxinus velutina 75 (velvet ash)

PALMACEAE (Palm Family)

    Sabal minor 314 (dwarf palmetto)

PHYTOLACCACEAE (Pokeweed Family)

    Phytolacca americana 77 (common pokeberry, pokeweed, poke)

    Rivina humilis 235 (pokeberry, inkberry, bloodberry rouge plant, pigeonberry)

PINACEAE (Pine Family)

    Abies religiosa 281 (fir)

    Cupressus benthamii 211 (cypress)

    Juniperus ashei 178 (ashe juniper, post cedar)

    Picea sitchensis 318 (spruce)

    Pinus aristata 339 (bristlecone pine)

    Pinus cembroides 125 (Mexican pinyon)

    Pinus edulis 8 (Colorado pinyon pine, pinyon pine, nut pine)

    Pinus ponderosa 333 (western yellow pine)

    Pinus strobus 3 (white pine)

    Pinus taeda 123 (loblolly pine, old field pine)

    Pseudotsuga taxifolia 316 (Britton douglas fir)

    Sequoia sempervirens 332 (giant sequoia)

    Tsuga mertensiana 338 (hemlock)

PLANTAGINACEAE (Plantain Family)

    Plantago rhodosperma 277 (redseed plantain)

    Plantago wrightiana 278 (wright plantain)

106

POLEMONIACEAE (Phlox Family)

Gilia incisa 37 (false flax, split gilia)

Loeslia ciliata 283

Phlox roemeriana 117 (roemer phlox)

POLYGONACEAE (Buckwheat Family)

Eriogonum annuum 13 (umbrella plant, annual wild buckwheat)

Polygonum convolvulus 304 (dull seed cornbind, cornbind weed, black bind weed,
wild buckwheat)

Polygonum punctatum 306

POLYPODIACEAE (Fern Family)

Adiantum capillus-veneris 209 (southern maiden's hair,
venus hairfern)

Asplenium resiliens 193 (blackstem spleenwort, little
ebony spleenwort)

Bommeria hispida 195 (hairy bommeria, dancing bommeria)

Cheilanthes lindheimeri 206 (fairy swords, lindheimer
lipfern)

Notholaena sinuata 114 (wavy cloak-fern, long cloak-
fern)

Pellaea atropurpurea 99 (purple cliff-brake, blue fern)

Phanerophlebia umbonata 203

Polypodium thyssanolepis 190

Woodsia mexicana 197 (Mexicana woodsia)

PORTULACACEAE (Purslane Family)

Portulaca pilosa 87 (moss rose, shaggy portulaca, hairy
portulaca, flowering moss)

PRIMULACEAE (Primrose Family)

    Samolus _cuneatus_ 263 (samolus cuneatus)

OSMUNDACEAE (Fern Family)

    Osmunda _regalis_ 317 (flowering fern)

RESEDACEAE (Mignonette Family)

    Resada _luteola_ 231

RHAMNACEAE (Buckthorn Family)

    Adolphia _infesta_ 257 (Texas adolphia)

    Ceanothus _greggii_ 260 (desert ceanothus, gregg
                      hornbrush)

    Colubrina _texensis_ 110 (hog plum, Texas
                      colubrina)

    Condalia _ericoides_ 83

    Karwinskia _humboldtiana_ 108 (coyotillo)

    Rhamnus _betulaefolia_ 265 (birch-leaf buckthorn)

    Zizyphus _mistol_ 266

ROSACEAE (Rose Family)

    Cerococarpus _breviflorus_ 232 (hairy mountain
                      mahogany, shaggy
                      mountain mahogany)

    Fallugia _paradoxa_ 233 (apache plume)

RUBIACEAE (Medder Family)

    Bouvardia _ternifolia_ 275 (scarlet bouvardia, trum-
                      petilla)

    Cephalanthus _occidentalis_ 270 (common buttonbush,
                      button willow)

    Houstonia _nigricans_ 35

RUTACEAE (Rue Family)

Ptelea trifoliata 246 (common hop-tree, stinking ash,
skunk bush, wafer ash, potato
chip bush)

Thamnosma texana 245 (dutchman's britches, Texas
desert rue)

SALICACEAE (Willow Family)

Populus deltoides 6 (eastern cottonwood, common
cottonwood, alamo cottonwood)

Populus grandidentata 93 (bigtooth aspen)

Populus monticola 97

Populus tacamahaca 89

Populus trichocarpa 102

Populus tremuloides 27 (quaking aspen, golden aspen,
alamo blanco)

Populus wislizeni 100 (Rio Grande cottonwood, alamo)

Salix gooddingii 236 (goodding willow, southwestern
black willow, bald-fruited
goodding willow)

SAPINDACEAE (Soapberry Family)

Sapindus drummondii 259 (wild chinaberry, western
soapberry)

Ungnadia speciosa 255 (Spanish or Mexican buckeye)

SAPOTACEAE (Sapodilla Family)

Bumelia lanuginosa 67 (gum elastic, shittim wood)

SCROPHULARIACEAE (Figwort Family)

Leucophyllum frutescens 111, 176 (ash plant, white
leaf, Texas silver-
leaf, barometer
plant, purple sage,
ceniza)

Maurandya antirrhiniflora 274 (snap dragon maurandya, vine blue snap dragon)

SELAGINELLACEAE (Selaginella Family)

Selaginella lepidophylla 104 (resurrection plant)

SOLANACEAE (Potato or Nightshade Family)

Chamaesaracha coronopus 273 (false nightshade, green false nightshade)

Lycium berlandieri 287 (berlandier wolfberry)

Nicotiana trigonophylla 81 (desert tobacco)

Solanum elaeagnifolium 107 (purple nightshade, silver leaf nightshade, bull nettle, white horse nettle)

STERCULIACEAE (Chocolate Family)

Ayenia pusilla 253 (dwarf ayenia)

Lythrum linearifolum 254

TILIACEAE (Linden Family)

Tilia americana 284 (linden, American basswood, whitewood)

TYPHACEAE (Cattail Family)

Typha domingensis 145 (narrow-leaf cattail)

Typha latifolia 237 (common cattail, reed mace, marsh pestle, water torch, bullsegg, candlewick, cat-o'-nine-tails)

ULMACEAE (Elm Family)

Celtis laevigata 69 (southern hackberry, sugar hackberry)

Celtis pallida 41 (granjeno, spiney hackberry, desert hackberry)

Ulmus crassifolia 80 (cedar elm)

UMBELLIFERAE (Parsley Family)

Aletes acaulis 264 (stemless aletes)

URTICACEAE (Nettle Family)

Parietaria pennsylvanica 238 (Pennsylvania pellitory)

VERBENACEAE (Vervain Family)

Aloysia lycioides 73 (whitebrush, whitebush, beebush)

Lantana macropoda 269 (veiny-leaf lantana, mejorana)

Lippia graveolens 109 (scented lippia, hierba dulce, redbrush)

Phyla incisa 268 (sawtooth frog-fruit, frog-fruit, spatulate-leaved frog-fruit, weighty frog-fruit, wedge-leaf frog-fruit)

Verbena canescens 105 (gray verbena)

VITACEAE (Grape Family)

Vitis rotundifolia 101 (muscadine grape, scruppernong)

ZYGOPHYLLACEAE (Calthrop Family)

Kallstroemia hirsutissima 21 (hairy caltrop)

Larrea divaricata 91 (creosote-bush, gobernadora)

Porlieria angustifolia 248 (guayacan)

A PRELIMINARY KEY TO SOME TEXAS POLLEN TYPES

John H. McAndrews

This key is to a series of references slides made during the summer of 1965 in the laboratory of Dr. Donald A. Larson, Department of Botany, The University of Texas. It is intended for limited use since it is only a preliminary guide to the identification of fossil pollen from Texas.

The nomenclature and format are patterned after Faegri and Iversen (1964). The measurements are usually of a single grain (the greatest dimension). Measurements of echinate grains do not include the spines.

1. VESICULATE, PINACEAE

    A. Body of grain > 100 u, cap 13 u thick at margins but thinner in middle . . . . . _Abies religiosa_ (281)*

    AA. Body < 90 u

        B. No constriction at junction of bladder and body, no marginal creat, bladder with fine internal reticulum, body 75 u . . . . _Picea sitchensis_ (318)

        BB. With constriction and marginal crest, bladder with coarser internal reticulum, body < 70 u . . . . _Pinus_ sp.

        C. Distally verrucate, "belly worts,"
           35 - 40 u . . . . . . _Pinus cembroides_ (125)
           48 - 50 u . . . . . . _P. edulis_ (8)

        CC. Distal verracae absent, 60 u . . . . _Pinus taeda_ (123)

2. POLYPLICATE, EPHEDRACEAE

    A. <u>Ca.</u> 15 straight "furrows," <u>ca.</u> 45 x 25 u . . . . _Ephedra aspera_ (241)
                               E. trifurca (240)
                                 E. torreyana (319)
                                 E. antisyphilitica (149)

    AA. <u>Ca.</u> 6 branches "furrows," <u>ca.</u> 50 x 22 u . . . . _Ephedra nevadensis_ (315)

---

*University of Texas Palynological Laboratory Accession Number.

3. <u>INAPERTURATE</u>

    A. Greater than 60 u

        B. Psilate, 90 u . . . . <u>Pseudotsuga</u> <u>taxifolia</u> (316)

        BB. Clavate, 75 u . . . . <u>Jatropha</u> <u>dioica</u> (243)

        BBB. Coarsely reticulate, 60 u . <u>Ruellia yuca-</u> <u>tana</u> (267)

    AA. Less than 60 u

        B. Scattered gemmae, 25 u . . . <u>Cupressus</u> <u>benthamii</u> (211)

        BB. Scabrate, 25 u . . . . <u>Populus</u> sp.

        BBB. Verrucate, 34 u . . . . <u>Aristolochia</u> <u>coryi</u> (63)

        BBBB. Rugulate, 45 u . . . . <u>Berberis tri-</u> <u>foliata</u> (55)

4. <u>MONOCOLPATE</u>

    A. Greater than 75 u, AMARYLLIDACEAE

        B. Ends of grain without reticulum, 120 u . . . . <u>Hymenocallis</u> <u>occidentalis</u> (33)

        BB. Reticulate throughout, 80 u. <u>Agave leche-</u> <u>guilla</u> (15)

    AA. 30-75 u

        B. Scabrate, 50-60 u . . . . . <u>Yucca baccata</u> (289) <u>Y. arkansana</u> (11)

        BB. Reticulate

            C. Reticulum finer near furrow, 40 u . . . . <u>Nolina texana</u> (49)

            CC. Reticulum uniform, 35 u <u>Sabal minor</u> (314)

    AAA. Reticulate, 25 u . . . . <u>Dasylirion leio-</u> <u>phyllum</u> (57)

5. <u>TRICHOTOMOCOLPATE</u> (no samples in collection)

6. <u>MONOPORATE</u>

    A.   Scabrate, annulate, GRAMINEAE,
           28 u . . . . <u>Andropogon</u> sp. (325)
           40 u . . . . <u>Stenotaphrum secundatum</u> (323)
           40 u . . . . <u>Tripsacum dactyloides</u> (322)

    AA. Finely reticulate, not annulate,
           25 u . . . . <u>Typha domingensis</u> (145)

7. <u>DICOLPATE</u> (no samples in collection)

8. <u>TRICOLPATE</u>

    A.   Psilate

        B.   Breaks on either side of "pore" area,
             26 u . . <u>Leucophyllum frutescens</u> (111,176)

        BB. Colpae geniculate,
             18 u ... <u>Quercus virginiana</u> (185)

    AA. Scabrate-verrucate

        B.   Small polar area, 25-30 u . . . .
             . . . . <u>Phytolacca americana</u> (77)

        BB. Large polar area

             C.   Sculpture fine,
                 27 u <u>Cercocarpus breviflorus</u> (232)

             CC. Sculpture coarser,
                 22 u <u>Quercus marilandica</u> (201)
                 25 u <u>Q. gricea</u> (161)
                 25-30 u . . . .
                 . . <u>Q. emoryi</u> (169)

   AAA. Baculate, spheroidal, 30-35 u

        B.   Thick exine . . . .
             . . . . <u>Dichondra brachypoda</u> (191)

        BB. Thin exine . . . .
             . . . . <u>Echinocactus texensis</u> (204)

  AAAA. Verrucate-spinulate, spheroidal,
             55 u . . <u>Echinocereus blanckii</u> (177)

AAAAA. Reticulate

    B. Gemmate-reticulate, short colpae,
        75 u . *Erodium texanum* (116)

    BB. Rugulo-reticulate, spherical, 35-40 u,
        CACTACEAE
                Mammillaria heyderi (184)
                Neomammillaria hebrichiara (194)

    BBB. Complete muri

        C. Operculate, fine reticulum, LEGUMINOSAE

            D. Prolate, large operculum, 38 x 22 u
                Dalea frutescens (208)

            DD. Oval, small operculum, 26 u
                Brogniartia sp. (220)

        CC. Not operculate

            D. Greater than 50 u, coarse reticulum,
                65 u
                Menodora heterophylla (106)

            DD. 25-30 u

                E. Fine reticulum or psilate with
                    breaks in furrow membrane, de-
                    fined "pore" area, 26 u
                Leucophyllum frutescens (176,111)

                EE. Medium reticulum, 30 u
                Acer grandidentatum (256)

            DDD. Spherical, fine reticulum, 17 u
                Drymaria debilis (234)
                Porlieria angustifolia (248)
                Halimolobus diffuscus (311)
                Sambucus coerulea (285)

9. STEPHANOCOLPATE

    A. Four colpae (sometimes 5 in *Fraxinus*), reticulate,
        large polar area

        B. Fine reticulum, 20 u
                Fraxinus texensis (65)
                F. velutina (75)

BB. Coarse reticulum with verrucas within muri,
35 u . . .Abronia angustifolia (221)

AA. Six colpae, psilate, 22 u
Phacelia congesta (113)

AAA. Eight colpae, echinae 4 u and with a hollow tip,
50 u . . . . Sicyos parviflorus (162)

10. PERICOLPATE

A. Echinate, spines small and blunt,
40 u . . . . Portulaca pilosa (87)

AA. Scabrate-reticulate, 30 u
Rivina humilis (235)

AAA. Verrucate

B. Perforate, spinulae (grains sometimes tricolpate) 55 u
Echinocereus blanckii (177)

BB. Angular, 4 copae, 20 u
Coldenia hispidissima (219)

AAAA.Rugulo-reticulate, with margo (grains sometimes
tricolpate) 35-40 u
Mammillaria heydeii (184)

11. DICOLPORATE (no samples in collection)

12. TRICOLPORATE

A. "Pore" an equatorial bridge or a ragged opening
in colpus membrane if bridge ruptured, colpus
often geniculate, < 35 u

B. Psilate

C. Angulaperturate, large polar area,
25 u Phoradendron coryae (175)
27 u Aloysia lydioides (73)
30 u Verbena canescens (105)

         CC. Not angulaperturate, small polar area,
              13 u _Koeberlinia spinosa_ (45)
              17 u _Buddleia marrubiifolia_ (292)
              20 u _Samolus cuneatus_ (263)
              30 u _Chamaesaracha coronopus_ (273)
              30 u _Diospyros texana_ (262)

BB. Scabrate

         C. Large polar area,
             19 u _Cocculus carolinus_ (230)

         CC. Small polar area,
             27 u _Cassia bauhinioides_ (229)

BBB. Striate,
         20 u . . ._Fallugis paradoxa_ (233)
         23 u . . ._Lycium berlandieri_ (287)
         27 u . . ._Nicotiana trigonophylla_ (81)

BBBB. Reticulate

         C.   Reticulum medium

             D.   Margo present

                 E.   Small polar area, 30-35 u
                 Fouquieria splendens (51)

                 EE. Greater than 25 u, 17 u
                 Salix gooddingii (236)
                        22 u
                 Maurandya antirrhiniflora (274)

             DD. Margo absent, 20 u
               Forestiera angustifolia (85)

         CC.   Reticulum fine

             D.   Apiculate, angulaperturate,
                slightly striate, 20 u
                Cleome gynandra (9)

             DD. Not apiculate, oval

                 E.   Slightly striate, 34 x 23 u
                 Mentzelia multiflora (43)

                    EE. Less than 25 u,
            15 u Sambucus coerulea (285)
            15 u Larrea divaricata (91)
            18 u Sophora secundiflora (216)
            19 u Cocculus carolinus (230)
            20 u Reseda luteola (231)
            20 u Lobelia brachypoda (286)
            20 u Schaefferia cuneifolia (252)
            22 u Cercis canadensis (250)
            25 u C. occidentalis (247)

AA. Pore circular

   B. Pore annulate

      C. Verrucate-reticulate

         D. Large polar area,
             26 u . . . .
             Sabatia campestris (120)

         DD. Small polar area,
              40 u . . . .
              Leucaena retusa (210)
              25 u . . . .
              Prosopis glandulosa (95)

      CC. Reticulate

         D. Fine reticulum, angulaperturate,
             small polar area, 26 u . . .
             Vitis rotudifolia (101)

         DD. Medium reticulum, 20 u . . .

             E. Small polar area . . . .
             Lonicera albiflora (276)

             EE. Large polar area, equatorial
             furrow . . . .
             Houstonia nigricans (35)

   BB. Pore not annulate

      C. Echinate, spines 1.5 u, pores indistinct,
          short spined, ca. 20 u, COMPOSITAE

D. Furrows long . . . .
   Iva ambrosiaefolia (180)

DD. Furrows short

   E. Long columellae . . . . .
      Iva texensis (170)
      I. frutescens (301)

   EE. Short columellae . . . .
      Ambrosia artemisiifolia (298)
      A. confertifolia (299)

CC. Psilate-scabrate

   D. Columellae distinct, prolate

      E. Sexine thickest at equator,
         30 x 20 u . . . .
         Eriogonum annum (77)

      EE. Sexine thickest near poles,
          23 x 16 u . . . .
          Phyllanthus polygonoides (242)

   DD. Columellae indistinct

      E. Pores acentric, 15 x 13 u
         Lappula redowskii (213)

      EE. Pores equatorial, 25 x 20 u
          Tournefortia hartwegiara (215)

CCC. Reticulate

   D. Peroblate, intruding vestibulae,
      37 u . . . .
      Tilia americana (284)

   DD. Not peroblate

      E. With 6 pseudocolpae, furrow
         operculate, 60 u . . . .
         Anisacanthus insignis (279)

      EE. No pseudocolpae

F. Apriculate, angulaperturate, fine reticulum, 26 u . . .
Vitis rotundifolia (101)

FF. Not apiculate, coarse reticulum, striate margo, 34 u . . . .
Eustoma exaltatum (295)

AAA. Pore equatorially elongated

B. Echinate, COMPOSITAE-(TUBULIFLORAE)

C. Spines > 2 u, 22 u
Zinnia acerosa (297)
27 u Lindheimera texana (300)

CC. Spines < 2 u

D. Spines 1.5 u, furrow long, 30 u
Flourensia cernau (147)

DD. Spines ca. 0.5 u, distinctly tectate, 30 u . . . .
Artemisia dracunculoides (310)

BB. Striate, RUTACEAE, 27 u . . . .
Ptelea trifoliata (246)
40 u . . Thamnosma texana (245)

BBB. Neither echinate or striate

C. Angulaperturate

D. Psilate, 20 u . . . .
Zizyphus mistol (266)
Rhamnus betulaefolia (265)

DD. Reticulate

E. Less than 30 u . . . .
Ceanothus greggii (260)
Condalia ericoides (83)
Colubrina texensis (110)
Karwinskia humboldtiana (108)
Ptelea trifoliata (246)

EE. Forty microns . . . .
Thomnosma texana (245)

CC. Not angulaperturate

   D. Reticulate

      E. Pore and furrow short, both of equal dimensions, spherical, 35 u . . . .
Garrya lindheimeri (293)

      EE. Not so

         F. Annulate pore in equatorial furrow, 22 u
Houstonia nigricans (35)

         FF. Not so, 27 u . . . . . .
Ptelea trifoliata (246)
                  30 u . . . . . .
Phyla incisa (268)
                  35 u . . . . . .
Ibervillea tenuisecta (171)
                  37 u . . . . . .
Lantana macropoda (269)

  DD. Psilate-scabrate

      E. Equatorial furrow present

         F. Exine thickest at poles, subprolate, 27 u . . . .
Polygonum convolvulus (304)

         FF. Exine uniform, spherical 25 u . . . .
Solanum eleagnifolium (107)

      EE. No equatorial furrow

         F. Tectate, exine thinning toward colpae, 20-30 u
Artemisia mexicana (305)
A. dracunculoides (310)
A. tridentata (321)

         FF. Not strongly tectate

            G. Large polar area

               H. Perprolate, equatorially constricted, 33 x 16 u . . . . .
Aletes acaulis (264)

HH. Not perprolate,
25 u . . . . .
Lippia graveolens (109)

        GG. Smaller polar area

           H. Thick exine, margo,
distinct columel-
lae, 40 u . .
Euphorbia mauritanica (244)

           HH. Thin exine, no
margo, indis-
tinct columellae,
30 u . . . . .
Diospyros texana (262)

13. STEPHANOCOLPORATE

A. Some colpae without pores

    B. Three of 12 colpae with pores, psilate, barrel-
shaped 22 x 16 u . . . .
. . . . Heliotropium angustifolium (222)

    BB. Three of 9 colpae with pores, striate, 22 u .
. . . . Heimia longipes (261)

    BBB. Three of 6 colpae with pores, striate, 30 u .
. . . . Lythrum linearifolium (254)

AA. All colpae with pores

    B. Six apertures, psilate, 23 x 19 u . . . . . .
. . . . Lithospermum multiflorum (212)

    BB. Four colporate, psilate-scabrate

        C. Pore an equtorial bridge, 20 u . . . . .
. . .Rhipsalis cassytha (187)

        CC. Pore equatorially elongated and protruding,
20 u Bumelia lanuginosa (67)

        CCC. Pore circular, pear-shaped, 25 x 20 u . .
. . Tournefortia hartwegiara (215)

14. PERICOLPORATE (no samples in collection)
Pericolporate types occasionally occur within species
which are normally tricolporate.

    A. Echinate, tectate, Tubliflorae . . . .

15. DIPORATE

    A. Annulate, irregular scabrae, 18 u . . . .
         . . . . Morus microphylla (59)

16. TRIPORATE

    A. Vestibulate, > 60 u, ONAGRACEAE

        B. Vestibulum with parallel stripes, psilate-
            scabrate

           C. Grain concave, 125 u . . . .
              . . . . . Oenothera speciosa (172)

           CC. Grain convex

                D. Vestibulum small, 75 u . . . .
                   . . Jussiaea diffusa (163)

                DD. Vestibulum large, > 79 u,
                   130 u Gaura coccinea (186)
                   100 u G. macrocarpa (202)

        BB. Vestibulum not striped but constricted at
            junction with grain

           C. Pore a slit, slightly striate, 105 u
              . . . . . Oenothera lampasana (182)

           CC. Pore round, baculate, 130 u . . . .
              . . . . . Oenothera albicaulis (168)

  AA. Less than 50 u

        B. Psilate

           C. Vestibulate, 21 u . . . .
              . . . . . Betula glandulosa (312)

CC. Not vestibulate

 D. Annulate

  E. Exine thin, 21 u . . . .
   <u>Maclura</u> <u>pomifera</u> (239)

  EE. Exine thicker, scabrate,
   23 u . . . .
   <u>Celtis</u> <u>pallida</u> (41)
   27 u . . . .
   <u>C.</u> <u>laevigata</u> (69)

  EEE. Exine thick, punctate, 30 u
   <u>Gilia</u> <u>incisa</u> (37)

 DD. Not annulate, slightly heteropolar,
  42 u <u>Carya</u> <u>texana</u> (205)

BB. Echinate, spines 4-5 u, slit-like furrow within
an annulus, 40 u . . . .
. . . . . . <u>Arbutilon</u> <u>incanum</u> (47, 173)
      <u>Sphaeralcea</u> <u>angustifolia</u>
          (39, 327)
      <u>S.</u> <u>coccinea</u> (328)
      <u>S.</u> <u>hastulate</u> (329)

BBB. Reticulate, annulate with ridges around
annulus, 25 u <u>Ayenia</u> <u>pusilla</u> (253)

17. <u>STEPHENOPORATE</u>

 A. Psilate

  B. Vestibulate, a pair of archi between each of
   4 or 5 pores, 25 u . . . .
   . . . . <u>Alnus</u> <u>crispa</u> (320)

  BB. Not vestibulate, annulate, 3-5 pores,
   23 u . . . . . <u>Celtis</u> <u>pallida</u> (41)
   27 u . . . . . <u>C.</u> <u>laevigata</u> (69)

 AA. Scabrate-punctate, 3-4 pores, annulate,
   30 u . . . . . <u>Gilia</u> <u>incisa</u> (37)

18. <u>PERIPORATE</u>

   A. Psilate

      B. Heteropolar, <u>ca</u>. 15 annulate pores,
32 u . . . . . <u>Juglans microcarpa</u> (61)

      BB. Not heteropolar, CHENOPODIACEAE and some
genera of AMARANTHACEAE

         C. Annulate, <u>ca</u>. 14 pores, 25 u . . .
. . . . . <u>Sarobatus vermiculatus</u> (308)

         CC. Not annulate, usually > 20 pores
. . . . . most genera of CHENOPODIACEAE

   AA. Rugulate-verrucate, <u>ca</u>. 12 pores with ragged margins, 55-70 u . . . <u>Opuntia leptocaulis</u> (188, 207)

   AAA. Verrucate, <u>ca</u>. 20 pores with annulae,
43 u . . . . <u>Loeslia ciliata</u> (283)

   AAAA. Reticulate

      B. <u>Ca</u>. 100 pores, high muri,
60 u . . . . . <u>Kallstroemia hirsutissima</u> (21)
<u>Allionia incarnata</u> (224)

      BB. Less than 30 pores, 60 u

         C. Reticulum fine, <u>ca</u>. 20 pores with 4-7
islets on membrane,
38 u . . <u>Liquidambar styracinflua</u> (307)

         CC. Reticulum coarse

            D. Verrucae in lumina, <u>ca</u>. 16 pores,
45 u <u>Polygonum punctatum</u> (306)

            DD. Lumina psilate, <u>ca</u>. 20 pores,
25 u <u>Phlox roemeriana</u> (117)

   AAAAA. Echinate

      B. More than 50 pores

         C. Spines < 5 u, NYCTAGINACEAE

            D. <u>Ca</u>. 50 pores, 140 u . . . . .
. . <u>Anulocaulis leisolenus</u> (226)

            DD.   Ca. 100 pores, 90 u . . . . . .
                  . . Oxybaphus albidus (29)

      CC.   Spines > 5 u

            D.   Spines acute and 18 u with baculae
                 in between, 50-100 pores,
                 65 u Hibiscus coulteri (189)

            DD.   Spines on convex base and 8 u,
                 ca. 150 pores, 80 u . . . . . .
                 . . Ipomoea costellata (167)

   BB.   Less than 40 pores

      C.   Annulate, ca. 12 pores, MALVACEAE

            D.   Pores in rows, 100 u . . . . . .
                 . . Gossypium hirsutum (17)

            DD.   Pores not in rows, 60 u . . .
                 . . Sida neomexicana (288)

      CC.   Not annulate

            D.   Thin exine, 6-12 pores, spines
                 5 u, 150 u . . . .
                 . . Cucurbita texana (181)
                 C. foetidissima (183)

            DD.   Thick exine, 12 pores,
                 NYCTAGINACEAE

                E.   Ca. 30 pores, spines 1-2 u,
                   120 u . . . .
                   Cyphomenis sp. (217)

                EE.   Ca. 15 pores, spines 4 u,
                   70 u . . . .
                   Boehaavia erecta (23)

19.   SYNCOLPATE

    A.   Ca. 8 pores in colpae, dice-like, 35 u . . . .
       . . . . Janusia gracilis (251)

AA. Pores absent

    B. Fingerprint striae, single encircling colpus,
33 u . . . . . . <u>Krameria glandulosa</u> (249)

    BB. Coarsely reticulate, several colpae
50 u . . . . . . <u>Chilopsis linearis</u> (115)

20. <u>HETEROCOLPATE</u> (no samples in collection)

21. <u>FENESTRATE, COMPOSITAE-LIGULIFLORAE</u>

    A. Echinate with 3-6 polar spines, with equatorial
ridge (Taraxacum type of Wodehouse, 1935)
30-35 u . . . . <u>Crepis petiolata</u> (325)
<u>Sonchus oleraceus</u> (343)

22. <u>DYADS</u> (no samples in collection)

23. <u>TETRADS</u>

    A. Persistent perisporium,
100 u . . . . <u>Selaginella lepidolphylla</u> (104)

    AA. No perisporium

        B. Psilate-scabrate,
40-50 u . . . . . <u>Arbutus menziesii</u> (290)
<u>A. texana</u> (294)

        BB. Spinulate,
50 u . . . . . . <u>Drosera annua</u> (309)

        BBB. Reticulate, monoporate,
40 u . . . . . . <u>Typha latifolia</u> (237)

24. <u>POLYADS</u>, LEGUMINOSAE

    A. Eight grains, not spherical

        B. One grain elongated, 150 x 90 u . . . . .
. . . . <u>Calliandra eriophylla</u> (228)

127

BB.   Seventeen x 14 u . .Mimosa biuncifera (128)

AA.   Sixteen grains, spherical,
40 u . . . . .Acacia greggii (174)

25.   MONOLETE, POLYPODIACEAE

A.   Grain 27 u, thin wrinkled perisporium
. . . . Phanerophlebia umbonata (203)

AA.   Grain > 35 u

B.   Perisporium wrinkled with high crests,
50 u . . . . . Polypodium thyssanolepis (190)

BB.   No distinct perisporium, grain scabrate,
38 u . . . . . Woodsia mexicana (197)

26.   TRILETE

A.   Perisporium fused with grain and verrucate-
regulate, spheroidal, 50 u, OSMUNDACEAE
. . . . Osmunda regalis (317)

AA.   Perisporium deciduous, grain angular, POLYPODIACEAE

B.   Perisporium wrinkled with high crests,
55 u . . . . . Pellaea atroupurpurea (99)

BB.   Perisporium verrucate, 45-60 u
. . . . . . Adiantum capillus-veneris (209)
Cheilanthes lindheimeri (206)
Notholaena sinuata (114)

BBB. Perisporium psilate to slightly wrinkled,
35 u . . . . . Bommeria hispida (195)

POLLEN ANALYSIS OF THE DEVIL'S MOUTH SITE

Vaughn M. Bryant, Jr.

INTRODUCTION

Pollen analysis is one of the major botanical methods employed in the reconstruction of past environments. Through careful and detailed study, palynologists are able to reconstruct to a large degree past vegetations and thereby obtain a key to past climatic conditions. The recognition of climatic conditions of the past often leads to greater insights into recorded cultural and physiographic changes. Pollen analysis became accepted as a research tool following Lennart von Post's demonstration of its potentialities in the analysis of Swedish peat bods during the early 1900's. Today the science of pollen analysis has effectively been utilized in such diverse fields as archeology, biogeography, geology, meteorology, petroleum exploration, paleobotany, and geochronology (Iversen, 1941; Anderson, 1955; Traverse, 1955; Callen, 1960; Martin and Sharrock, 1964; Kapp, 1965).

The practice of pollen analysis is possible because: 1) many plants emit great quantities of pollen or spores, 2) most pollen grains and spores have a chemically stable outer wall (exine) which resists deterioration, 3) exine structure is consistent within a species and is generally diverse among unrelated taxa, 4) inherited diagnostic features allow pollen of one genus or species to be distinguished from types produced by other plant taxa, 5)the preserved pollen rain generally provides a reasonably accurate image of the regional anemophilous (wind-pollinated) vegetation. However, consideration of contemporary pollen rains is an important factor since not all pollen follows the same distributional pattern. Pollen grains from anemophilous plants such as pine and _Ephedra_ are buoyant and can travel great distances in air currents. Other less buoyant anemophilous pollen types such as _Carya_ and _Zea_ are only carried short distances from their source. The distribution of pollen from zoophilous (insect-pollinated) plants is even more restricted.

Through the use of proper laboratory procedures it is possible to remove the unwanted matrix surrounding the resistant pollen grains (_i.e._, rock, coal, peat, soil) and to concentrate the pollen thereby permitting a microscopic analysis of the grains. A statistically significant number of pollen grains are then counted and recorded on tabular sheets as to their family and genus (such as

129

Table 3). When this is completed all of the individual
analyses are generally presented visually in the form of
a continuous pollen diagram in which the percentages of
each pollen type within a given level are plotted against
the other types (for example, Figures 21 and 22). Generally
the time span (or succession of strata) is plotted on the
vertical scale while comparative percentages of pollen at
each level are plotted horizontally. The palynologist
then studies the appearance, disappearance, and fluctuations in the percentages of pollen types on the pollen
record and attempts to reconstruct the environmental
conditions during the deposition of each soil stratum.

During a pollen analysis care must be taken not to
"overuse" the pollen evidence in trying to reconstruct
a quantitative distribution of vegetation types. The
number of pollen grains that each different plant type
contributes to the pollen record depends upon three major
factors: 1) the abundance and position of the plant in
relation to the total vegetation cover, 2) the inherent
pollen productivity capability of each plant species, and
3) the method of pollen dispersal (Davis, 1963). The following example illustrates how the pollen rain does not
always present accurate quantitative data concerning the
percentages of each plant type in the total vegetation
cover. The percentages of pollen deposited upon the surface of a forest containing 90% maple trees and 10% pine
trees would probably not reflect these percentages since
each pine male cone produces approximately 1,500,000
pollen grains while each maple flower only produces an
estimated 8,000 pollen grains (Gray and Smith, 1962).
Therefore, each pine male cone contains the equivalent
pollen of 188 maple flowers. A pollen diagram only in-
directly reflects the percentages of individual plant
types growing in a region during the periods of deposition.
If a soil sample contains 30% of pollen type "A" and 70%
of pollen type "B" it does not necessarily mean that 30%
of the total vegetation cover was composed of plants depositing pollen type "A" and that the remaining 70% of the
vegetation cover consisted of plants depositing pollen
type "B."

Professor Johs. Iversen of Denmark was one of the
first palynologists to apply the principles of pollen
analysis to the field of archeology (Iversen, 1941). In
his analysis of several Danish archeological sites he
showed how the pollen record documented Neolithic man's
dominance over his local environment. Iversen noticed

TABLE 3. POLLEN TABULATION SHEET

Archeological Site_____

Stratum or level_____   Analyst_____

Sample number_____   Total Count_____

Pollen Type	Total	Percentage
Compositae		
Low spine		
High spine		
Artemisia		
Liguliflorae		
Total Compositae		
Gramineae		
Pinus		
Juniperus		
Quercus		
Prosopis		
Celtis		
Cheno-Ams		
Cactaceae		
Mammillaria		
Opuntia		
Nyctaginaceae		
Cyperaceae		
Acacia/Mimosa		
Ephedra		
torreyana type		
nevadensis type		
Euphorbiaceae		
Malvaceae		
Agave		
. . . .		
. . . .		
. . . .		
. . . .		
Fungal spores		
Fern spores		
Selaginella		

131

that the pollen record from deposits dating from the
transitional period between the Atlantic and Sub-Boreal
periods showed a sudden decline in the percentages of
tree pollen and a sharp increase in herb and shrub pollen. This change in the pollen record was accompanied by
a thin layer of charcoal and the gradual increase in the
percentages of pollen from cultivated plants. Thus Iversen was able to infer that there was pollen evidence
indicating that Neolithic man introduced agriculture to
Northern Denmark. The sudden decline in the percentage
of tree pollen followed by a charcoal layer and an increase in the percentages of cereal pollen is evidence
that Neolithic man cleared the forests using the slash
and burn method and then cultivated the cleared land
with cereals.

Today the potentials of pollen analysis in the field
of archeology are seemingly unlimited. Thus far pollen
analyses have provided archeologists with a means of relative dating, a method of determining past environmental
conditions, an insight into the diets and practices of
ancient cultures through palynological studies of human
coprolites, and an indication of how prehistoric man
adapted to the conditions of his environment. However, a
pollen analysis of a single site rarely provides the
answers to every paleoecological question. Even under
ideal conditions a complete pollen analysis of a single
site can only offer limited ecological and anthropological
data. On the other hand, as more analyses are conducted
in a specific area of investigation (for instance, the
Amistad Reservoir area), more paleoecological and anthro-
pologically useful information becomes available
for interpretation.

The first pollen analysis of an archeological site
in the Amistad Reservoir area was attempted by LeRoy
Johnson, Jr., in 1963 (Johnson, 1963). His preliminary
studies of pollen samples from two rockshelters (Centipede
and Damp Caves) demonstrated that pollen analyses were
applicable as a research tool in reconstructing the late
Quaternary environment of this arid region of southwest
Texas. In 1965, Drs. Donald A. Larson, Richard H. Hevly,
John H. McAndrews, and the author collected soil samples
from four additional archeological sites in the Amistad
Reservoir area (Bonfire Shelter, Devil's Mouth Site,
Devils Rockshelter, and Eagle Cave) and conducted a pollen
analysis of each site. This report embodies the results
of one of these sites, the Devil's Mouth Site (41 VV 188).

SITE GEOGRAPHY AND DESCRIPTION

The area west of Del Rio, Texas, in the watershed of the Rio Grande, Devils, and Pecos rivers offers an excellent opportunity for archeological investigation. Previous excavations in this area have provided evidence that primitive man inhabited this region for at least ten to twelve thousand years (Epstein, 1963; Johnson, 1964; Dibble, 1965; Nunley et al., 1965; Ross, 1965). Unfortunately, many of the more than 300 known archeological sites in this region will be flooded by the end of 1969, when the Amistad International Dam is completed and its reservoir is filled to floodpool level (Figure 3).

The Devil's Mouth Site is located at the confluence of the Rio Grande and Devils River and is one of these archeological sites that will be inundated by the Amistad Reservoir. The site lies on an alluvial terrace approximately fifty feet above the present water level and is comparatively flat except for the steeply eroded edges along the water's edges. North of the terrace lies a long limestone ridge that runs the full length of the terrace and juts out almost to the water's edge at the Devils River (Figure 11).

The Devil's Mouth Site was discovered during the original archeological survey of the Amistad Reservoir in 1958 (Graham and Davis, 1958). In December of 1959, the site was tested by a field crew from the Texas Archeological Salvage Project and found to contain sufficient cultural deposits to warrant further excavation. In September of 1961, another field crew, under the direction of LeRoy Johnson, Jr., returned to the terrace and carried out further excavations at the Devil's Mouth Site.

The largest excavation pit at the Devil's Mouth Site reached a maximum depth of thirty-six feet and uncovered twenty-four recognizable strata (Johnson, 1964). Stratigraphically, the Devil's Mouth Site can conveniently be divided into two broad units separated by an erosional surface. The lowest unit contains one midden zone and ten alternating strata of silt, sand, and clay. The upper stratigraphic unit is composed of thirteen strata of alternating midden layers and sterile sand deposits (Figure 12).

One of the problems associated with the antiquity of the Devil's Mouth Site is the absence of radiocarbon dates. Johnson was able to construct a relative time sequence for the Devil's Mouth Site through comparisons of projectile

134

points and natural stratigraphy with similar sites in the
Amistad and Central Texas regions. According to his
chronology he calculated the following time periods for
the Devil's Mouth Site: 1) Paleo-Indian horizon, approximately 6,000-5,000 B.C.; 2) Early Archaic, 4,500-3,000
B.C.; 3) Middle Archaic, 3,000-1,500 B.C.; 4) Late
Archaic, 1,500 B.C. - A.D. 1,500 (Table 4).

Recently, Dr. Dee Ann Story has re-evaluated the time
periods for many of the archeological sites in the Amistad
Reservoir area. Her classification system divides the
late Quaternary into eight periods beginning with Period I
(greater than 7,000 B.C.) and ending with Period VIII
(1,600 A.D.-present). In this report all time references
refer to Story's classification system unless stated
otherwise (see archeological background section).

MATERIALS AND METHODS

Fossil Pollen Samples

On March 4, 1962, LeRoy Johnson, Jr., John Greer, and
Mark Parsons collected a series of twenty-eight soil samples which were set aside for future palynological studies
of the Devil's Mouth Site. Three of the samples were collected in the gravel and overlying sand deposits in Area C
of the site (Figure 11). The remaining twenty-five samples
were collected in Area A of the site by forcing the open
ends of one-inch wide steel pipes into the cleaned vertical
walls of the largest excavation pit. Each pipe was then
sealed with rubber stoppers, labeled, and stored for
future use (Table 5).

On several occasions during 1965, the writer collected
additional columns of samples from the exposed profiles at
the Devil's Mouth Site. However, these more recently collected samples were never used since their exact provenience
in relation to the original excavations could not be determined. After nearly four years of exposure to the elements,
the excavation pits had eroded and partially filled with
sediments. This made correlations with previously drawn
profile sheets difficult, and, in most cases, impossible.

Modern Pollen Samples

Three modern soil samples were collected and analyzed
during the course of this study. On March 4, 1962, LeRoy
Johnson, Jr., and Mark Parsons collected a few ounces of

TABLE 4. DEVIL'S MOUTH SITE TIME CHRONOLOGY

Stratum	Johnson's Sequence	Story's Sequence
1		Periods VI & VII
2		(200 B.C. - A.D. 1600)
3	1,000 A.D.	
4		
5		
6	-0-	
7		
8	1,000 B.C.	Period V
9		(1,000-200 B.C.)
10	2,000 B.C.	
11		Period IV
12	3,000 B.C.	(2,500-1,000 B.C.)
13		Period III
14		(4,000-2,500 B.C.)
15		
16	4,000 B.C.	
17		
18		Period II
19		(7,000-4,000 B.C.)
20	5,000 B.C.	
21		
22	6,000 B.C.	Period I
23		(Greater than 7,000 B.C.)
24		
Gravels in Area C		

TABLE 5. SOIL SAMPLES FROM THE DEVIL'S MOUTH SITE

According to the original field notes the provenience of each of the twenty-eight soil samples used in this study is as follows:

Samples 1, 2, and 3
    These three samples were collected from the southern wall of the western extension of Test Excavation 4 in Area C of the Devil's Mouth Site. Sample 1 was taken from the lower gravel deposits beneath a large bone, probably elephant. Sample 2 was collected from the upper gravel deposits three-tenths of a foot from the top of the gravel zone. Sample 3 was taken one foot above the gravel in the overlying clay and sand deposits.

Sample	Depth below surface	Stratum	Grid Coordinates
4	.5 feet	1	N400/W400
5	1.0 "	2	"
6	1.5 "	3	"
7	2.5 "	4	N400/W397.5
8	3.0 "	5	"
9	3.5 "	6	"
10	4.5 "	7	"
11	5.5 "	8	N400/W403
12	6.5 "	9	"
13	7.0 "	10	N395/W400
14	8.2 "	11	"
15	8.7 "	12	"
16	9.2 "	13	"
17	9.7 "	14	"
18	10.2 "	15	"
19	11.2 "	16	"
20	12.2 "	17	"
21	12.7 "	18	"
22	13.7 "	19	N390/W403
23	15.7 "	20	"
24	17.7 "	21	"
25	19.7 "	22	"
26	21.7 "	22	"
27	23.7 "	23	N392.5/W403
28	25.7 "	24	"

sediment from the bottom of a four-foot high stone-lined cattle tank three miles north of the Devil's Mouth Site on the Figueroa Ranch. The remaining two surface samples were collected by the author at the Devil's Mouth Site in December of 1965. One of the samples consisted of 30-40 pinches of surface dirt collected over an area fifty meters square surrounding the excavations at the Devil's Mouth Site. The other surface sample collected in 1965 consisted of approximately 20 pinches of surface dirt collected along the top of the limestone ridge 150 feet northwest of the site.

Extraction Techniques (see Appendix B for details)

The extraction techniques developed for processing these samples were a combination of several accepted methods. First, the samples were treated with hydrochloric acid to remove the calcium carbonates; then they were placed in a heated solution of 10% potassium hydroxide to remove carbon and to soften cellulose tissue. When this was completed, the samples were treated with hydrofluoric acid to remove silicates and then acetylated to remove the unwanted cellulose and other organic compounds.

Mounting (see Appendix C for details)

After soil digestion, the concentrated pollen residue from each sample was transferred to five milliliter shell vials containing a few drops of 2000 cs silicone oil and thoroughly mixed with the oil to insure statistically accurate samples. When this procedure was completed, portions of the samples were then transferred to one by three inch glass microscope slides, covered with number 1 cover slips, and permanently mounted for microscopic analysis.

Observations

Observations for this study were made with an A.O. Spencer phase-contrast microscope having lenses of 10X, 20X, 43X, and 97X and ocular lenses of 15X. Most of the samples were examined at a magnification of either 645 or 1,455 diameters.

Pollen Reference Collection

Beginning in the summer of 1965 and continuing through to the present, pollen samples for a reference collection have been obtained and processed. Approximately 350 pollen samples have been collected from The University of Texas Herbarium (see paper on Amistad pollen reference collection). Most of the pollen selections were of plant types known to exist either in the Amistad Reservoir area or within a hundred-mile radius of the reservoir area (Blair, 1950; McDougall and Sperry, 1951; Gould, 1962). Other selections such as _Picea_, _Pseudotsuga_, _Ephedra nevadensis_, and some species of _Pinus_ were chosen because they are anemophilous types which do not live in the Amistad region but appear in the fossil pollen record of the area. The herbarium samples were processed in heated 10% potassium hydroxide followed by acetylation. Each processed sample was stained with safranin and mounted in 2000 cs silicone oil. After all the samples were processed, a pollen key was prepared for the reference collection (McAndrews, herein). The Amistad Pollen Reference Collection and the accompanying pollen key are currently being kept in the Botanical Palynological Laboratory of The University of Texas.

Identification of pollen types in this report are based upon comparative studies with pollen samples in both the Amistad Reference Collection and in other general reference collections previously prepared by the Botanical Palynological Laboratory.

Pollen Counts

A standard 200-grain count was reached in all but six of the processed samples used in this study. Each identifiable whole pollen grain was identified and recorded on tabular sheets (Table 3). The only exceptions were clusters of Compositae pollen easily recognizable as coming from broken anthers. These clusters were counted as single grains. Fragments of known pollen grain types were also included in the standard count. Broken _Pinus_ grains, for example, were counted as one-third of a grain for each bladder and one-third for the body of the grain. At the end of each standard count all fractions were rounded off to the nearest whole number. Badly crushed and deteriorated grains beyond identification were excluded from the pollen counts.

The primary count, or first pollen count, was a standard 200-grain count which included all pollen types. A second count of 150 pollen grains included all pollen types except Compositae. The second count was reached by using all of the non-Compositae pollen grains from the first count and then counting additional non-Compositae grains from newly prepared slides. For example, in the first count of the fossil material from Stratum 3, 121 of the total 200 pollen grains were Compositae. The second count of 150 grains was reached by adding the 79 non-Compositae pollen grains from the primary count of Stratum 3 to 71 additional non-Compositae grains counted on other slides from the same sample.

In this report all reference to "primary" or "first" count refers to the 200-grain count which included all pollen types. Reference to the "second" or "secondary" count refers to the 150 grain count which excluded Compositae but included all other pollen types.

RESULTS

During the course of analyzing all pollen samples, thirty-three pollen types representing twenty-four families were encountered (Table 6). Twelve of these types (Alnus, Celtis, Ephedra nevadensis, Euphorbiaceae, Gaura, Jussiaea, Liguliflorae, Liliaceae, Liquidambar, Maclura, Mammillaria, Typha) were found only in the fossil record while Boraginaceae and Salix pollen were found only in the modern samples. In general the results are best represented by the completed pollen diagrams (Figures 21 and 22).

Fossil Pollen Record

All of the originally collected soil samples from the Devil's Mouth Site were analyzed for the presence of pollen. Nineteen of the twenty-eight samples contained sufficient pollen for a standard 200-grain count. Three of the other nine samples contained between 92-134 pollen grains and were therefore included in the report as partial counts. The remaining six soil samples contained insufficient pollen for even a partial analysis and were therefore excluded from the pollen diagram. Unfortunately, the six non-productive samples were three samples from the gravels in Area C of the site and the three deepest alluvial samples from Area A.

FIGURE 21. Primary pollen diagram, Devil's Mouth Site.

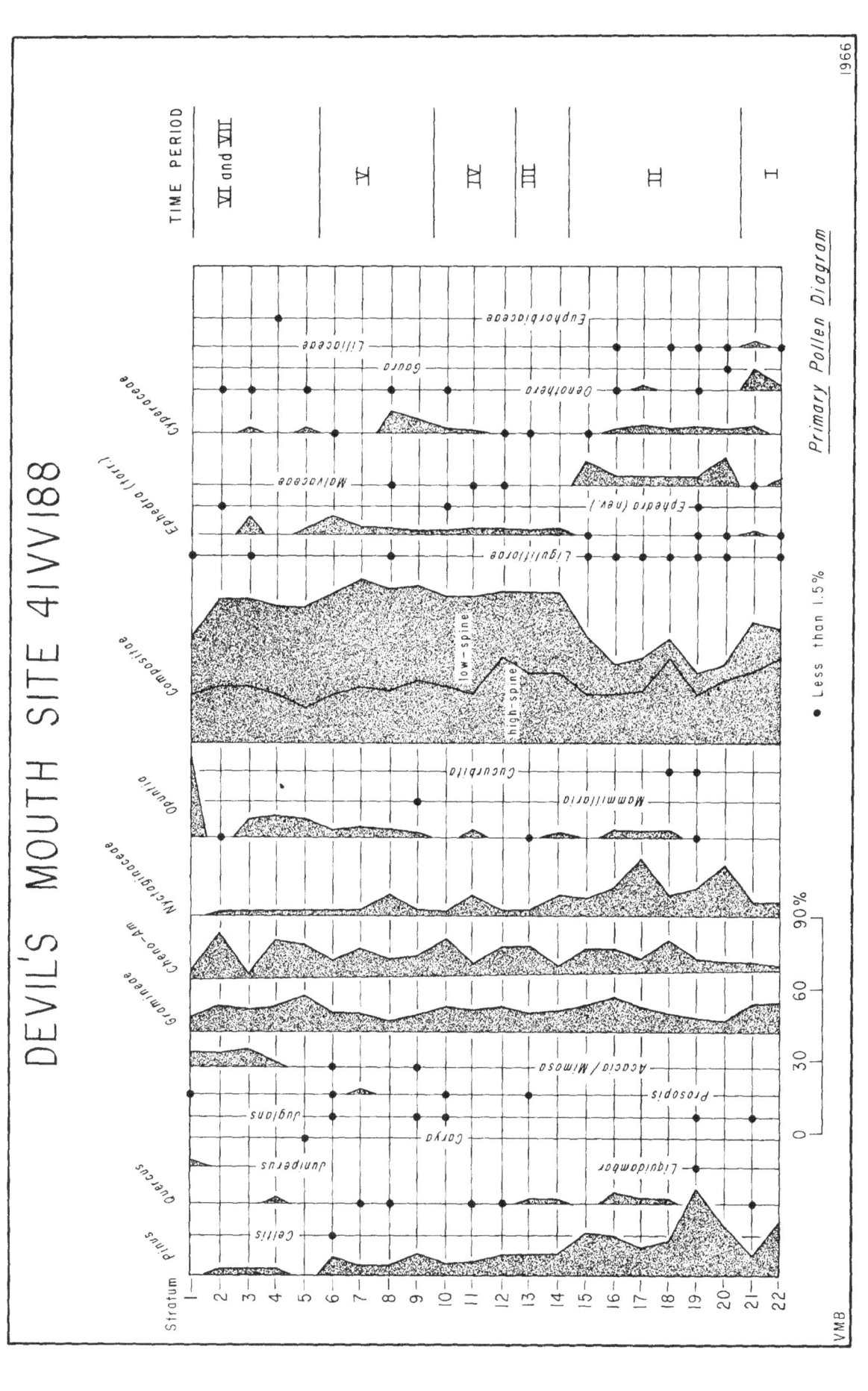

DEVIL'S MOUTH SITE 41VV188

Primary Pollen Diagram

• Less than 1.5%

1966

VMB

FIGURE 22. Secondary pollen diagram, Devil's Mouth Site,
including percentages of arboreal pollen
from both the first and second count.

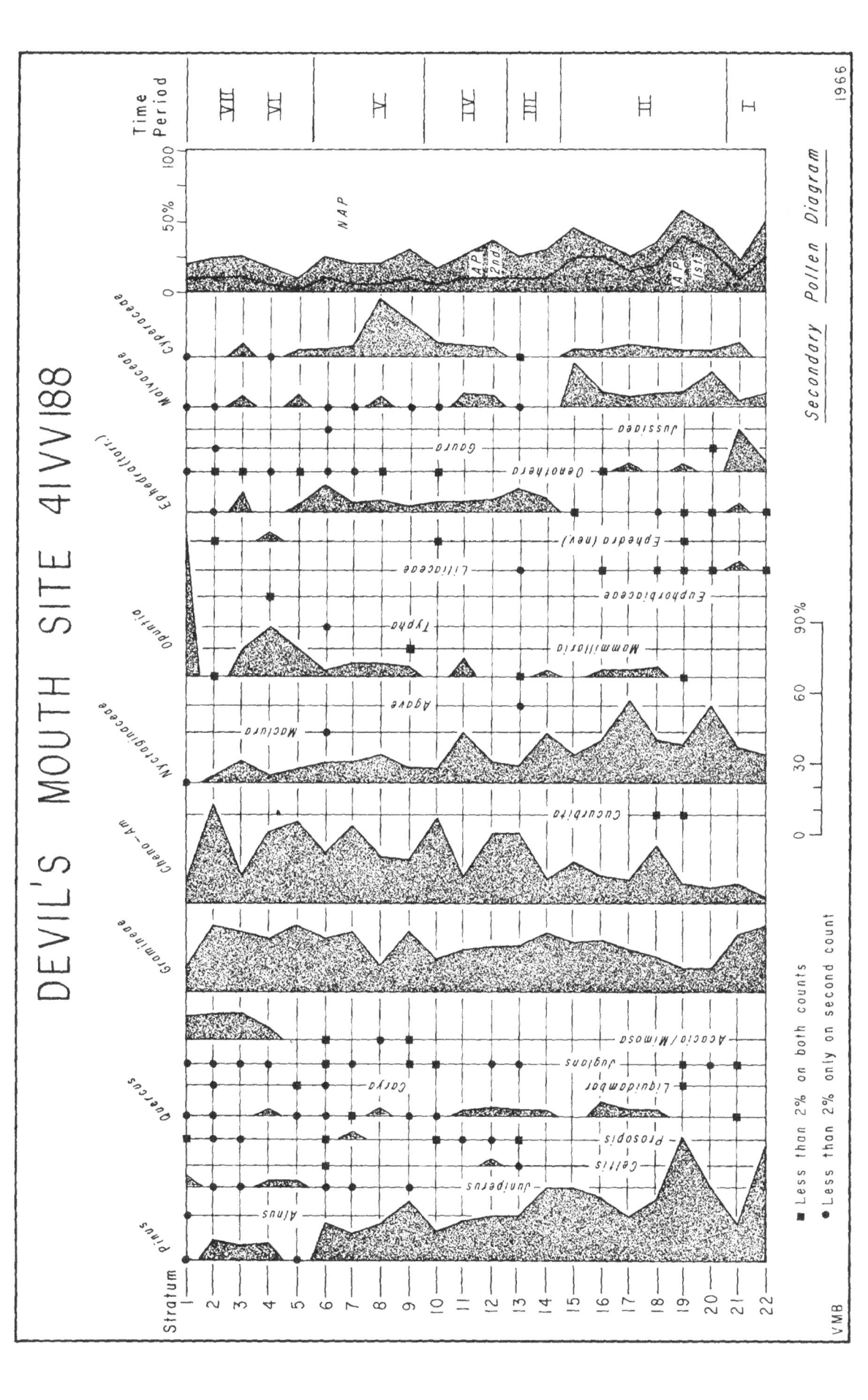

DEVIL'S MOUTH SITE 4IVV188

Secondary _Pollen Diagram_

1966

■ Less than 2% on both counts

● Less than 2% only on second count

VMB

Primary Pollen Diagram

A study of the primary pollen diagram is basically a study of the fossil Compositae pollen at the Devil's Mouth Site (Figure 21). Pollen of the Liguliflorae, high-spine Tubiflorae, and low-spine Tubiflorae types comprise between 30-67% of the total pollen in each of the fossil soil samples.

Low-spine Compositae pollen accounts for more than one-half of the total Compositae pollen in each of the samples from the upper fifteen strata. In a few of these strata (see primary pollen diagram) the dominance of low-spine Compositae grains reaches a peak of nearly three to one while in other strata the ratio of low-spine grains to other Compositae types is almost equal. These ratios would be even higher if each grain of the low-spine Compositae clusters had been counted individually instead of treating them as single grains. In the remaining seven strata the high-spine pollen types dominate the Compositae spectra.

Secondary Pollen Diagram

The dominance of a single pollen type (Compositae) in all of the samples obscured and prevented an accurate evaluation of other pollen types within each strata. To correct this overrepresentation a second pollen count was made for each stratum. The second count excluded all of the Compositae grains but included all other pollen types. The secondary pollen diagram shows the results of the second count (Figure 22).

As seen in the diagrams, the percentage of Cheno-Am pollen fluctuated very little from sample to sample in the results of the primary count but during the second count these fluctuations became much more pronounced. Many of the other pollen types such as Pinus, Acacia, Mimosa, Nyctaginaceae, Opuntia, and Oenothera show fairly uniform increases on the second pollen count. Several pollen types, specifically Alnus, Agave, Jussiaea, Maclura, and Typha were encountered only during the second count.

Modern Pollen Rain

The three modern surface samples yielded sufficient pollen for a standard 200-grain count even though a number of the grains were badly deteriorated. The sampled contained twenty-two different pollen types and several genera

140

of Polypodiaceae along with two species of Selaginella, all of which were identified but excluded from the standard counts. In addition to the identified types, approximately twelve pollen grains from each of the surface samples were of unknown types which did not compare morphologically with any of the pollen types in either of the reference collections used in this study.

Most of the pollen types in the three modern surface samples are both arboreal and anemophilous. In the surface sample from the terrace 47% of the pollen is from arboreal plants and 65% is anemophilous. Both of the other surface samples reflect a similar pattern. The rim surface sample contained 67% arboreal and 79% anemophilous pollen while the stock tank sample contains 57% arboreal pollen and 64% anemophilous pollen (Figure 23).

Each of the surface samples contains significant percentages of Pinus, Compositae, Carya, Gramineae, Cheno-Ams, and Acacia/Mimosa pollen. Certain other pollen types such as Nyctaginaceae, Cyperaceae, Oenothera, and Opuntia are found in the surface samples but are weakly represented.

Pollen Types

The Compositae pollen in this report is divided into three major categories based upon their morphological differences: 1) Liguliflorae, 2) high-spine Tubiflorae, and 3) low-spine Tubiflorae. Artemisia was not assigned to a separate category since only five of the more than 6,000 fossil grains examined in this study were considered to be definitely of the Artemisia type. Artemisia was also weakly represented in the three surface samples and therefore omitted as a separate category in those samples as well. The grains belonging to the Liguliflorae group of the Compositaes are fenestrate and easily distinguishable from the other types of Compositae (Wodehouse, 1935). The other two categories, high-spine and low-spine, are arbitrary divisions of the subfamily Tubiflorae. The high-spine group is generally insect or self-pollinated and their echinate pollen grains have spines greater than two microns in length (Martin, 1963). The low-spine group, on the other hand, is often anemophilous, having spines less than two microns in length.

In 1963, Paul S. Martin proposed the term "Cheno-Ams" for pollen of the genera Atriplex, and Chenopodium in the

TABLE 6.

POLLEN TYPES FOUND IN THE SOIL SAMPLES FROM
THE DEVIL'S MOUTH SITE

AMARYLLIDACEAE
     Agave sp.

BETULACEAE
     Alnus sp.

BORAGINACEAE

CACTACEAE
     Mammillaria sp.
     Opuntia sp.

CHENO-AM
     (Atriplex sp., Amaranthus sp., and Chenopodium sp.)

COMPOSITAE
     Low-spine types
     High-spine types
     Liguliflorae types

CUCURBITACEAE
     Cucurbita sp.

CYPERACEAE

EPHEDRACEAE/GNETACEAE
     Ephedra (nevadensis type)
     Ephedra (torreyana type)

EUPHORBIACEAE

FAGACEAE
     Quercus sp.

GRAMINEAE

HAMAMELIDACEAE
     Liquidambar sp.

JUGLANDACEAE
     Carya sp.
     Juglans sp.

142

TABLE 6 (Cont'd)

LEGUMINOSAE
  Acacia sp./Mimosa sp.
  Prosopis sp.

LILIACEAE

MALVACEAE

MORACEAE
  Maclura sp.

NYCTAGINACEAE

ONAGRACEAE
  Gaura sp.
  Jussiaea sp.
  Oenothera sp.

PINACEAE
  Juniperus sp.
  Pinus sp.

SALICACEAE
  Salix sp.

TYPHACEAE
  Typha sp.

ULMACEAE
  Celtis sp.

FIGURE 23. Modern pollen rain, Figueroa ranch.

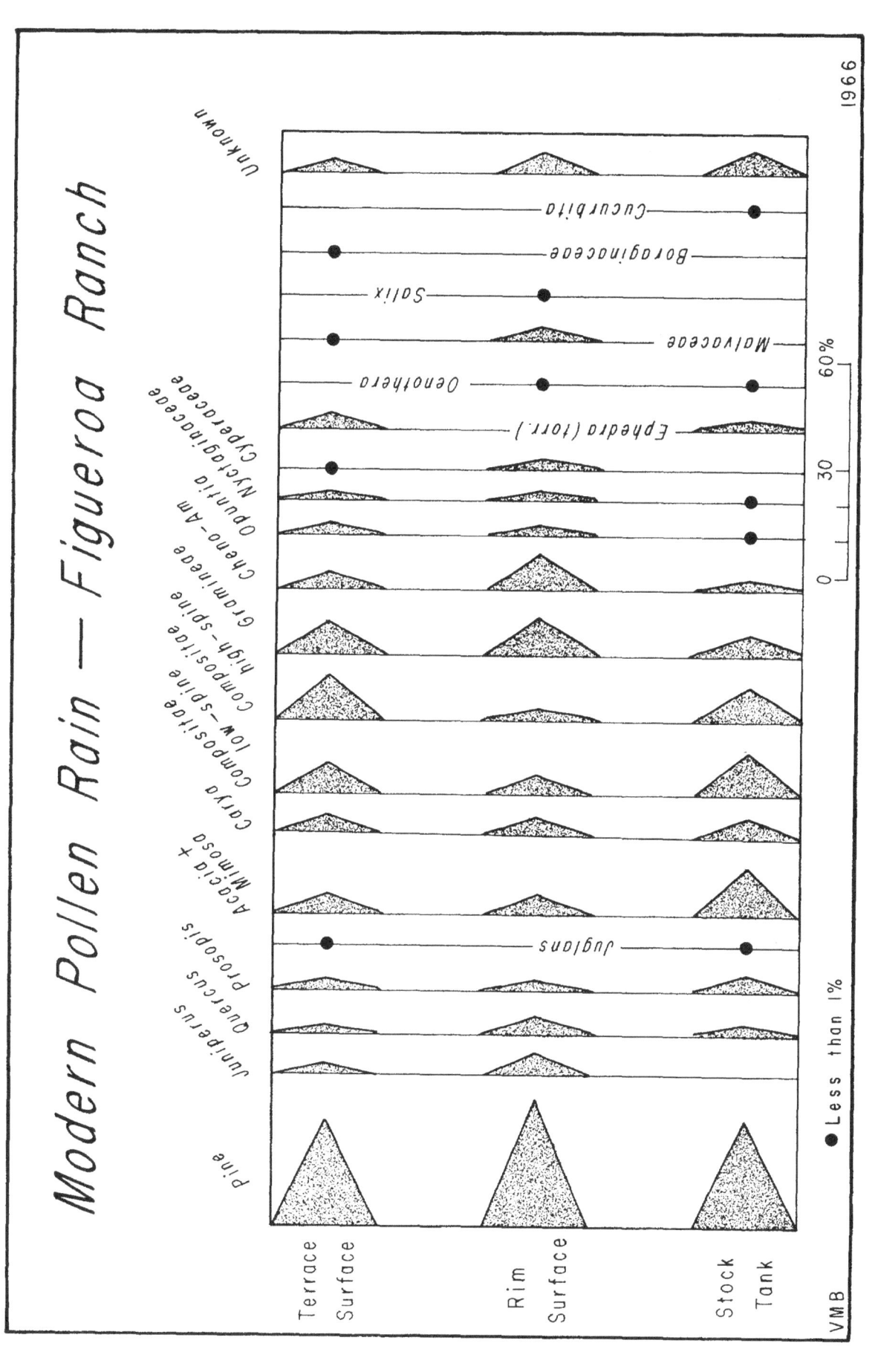

Modern Pollen Rain — Figueroa Ranch

Chenopodiaceae, and the genus Amaranthus in the family Amaranthaceae (Martin, 1963). These genera were combined into this single artificial group termed Cheno-Ams since their pollen grains are so morphologically similar that it is nearly impossible to distinguish one type from another.

The Ephedra pollen identified in this report is divided into two groups, Ephedra nevadensis and Ephedra torreyana (Steeves and Barghoorn, 1959). The Ephedra torreyana group is distinguished by its large number (ca. 15) of unbranched straight furrows while the Ephedra nevadensis group tends to have fewer furrows (ca. 6) with undulating ridges and a number of conspicuous hyaline strands intersecting the furrows at various points along the grain. In this report Ephedra antisyphilitica and Ephedra torreyana are placed in the Ephedra torreyana group while Ephedra nevadensis and Ephedra aspera are placed in the Ephedra nevadensis group.

The Cactaceae pollen in this report is divided into two groups, Opuntia and Mammillaria. The Opuntia grains are periporate and probably belong to the species Opuntia lindheimeri. The Mammillaria pollen type is tricolpate and measures 45 microns in diameter (Tsukada, 1964).

Most of the fossil Malvaceae grains are similar to pollen grains of the genus Sphaeralcea. The grains are echinate, triporate, annulate, and have a body diameter of 35-40 microns. The surface structuring on most of the grains is badly deteriorated thereby preventing a positive identification as to genus.

The Nyctaginaceae pollen was not subdivided into genera since it consisted of several unknown types which did not compare morphologically with any of the genera available for comparison in the Palynological Laboratory.

All of the large Gramineae pollen grains were measured and found to be smaller than thirty-five microns. Furthermore, none of them compared morphologically with reference material of either Tripsacum, Teosinte, or Zea (Irwin and Barghoorn, 1965; Whitehead, 1965).

DISCUSSION

The pollen record from the Devil's Mouth Site is significant from several points of view: 1) it demonstrates a general progress towards aridity following the last full-glacial period, 2) it provides insights into the cultural

144

history of prehistoric and historic man, and 3) it offers
additional data for use in the understanding of the
Quaternary period and especially the segment of this
period known as the Altithermal (see Appendix A).

In the Southwest, climatic conditions during the
"so-called" Altithermal period have become a subject of
both investigation and disagreement among many geologists
and palynologists. Ernst Antevs, Alan L. Bryan, and others
claim that the Altithermal Period (ca. 5,500-2,500 B.C.)
in the Southwest was dry and warm. They claim that the
severe arroyo erosion and calichication which occurred
during this period was caused by flash flooding resulting
from an arid climatic condition (Antevs, 1962; Bryan and
Gruhn, 1964). Contrary to this view, palynologist Paul
S. Martin and his associates believe that in the Southwest the climatic conditions during the Altithermal period
were moist and subpluvial. They argue that the severe
erosion and large alluvial deposits that occurred during
this period resulted from flooding caused by intense summer rains (Martin, 1963).

The pollen record at the Devil's Mouth Site
indicates that during part of Periods I, II, and III
(Altithermal deposits make up a portion of the sediments
of Periods II and III) the climatic conditions in the
Amistad region were more mesic (moist) than they are
today. This is evidenced by: 1) high pollen percentages
from plant types that tend to survive best in a mesic
environment, 2) low percentages of pollen from xerophytic
(drought resistant) plants, and 3) rapid alluvial build-
up on the terraces along the major rivers in the
Amistad region.

One of the major pollen contributors during Periods I,
II, and III is Pinus. In the upper deposits of Period I
and the lower deposits of Period II, the percentage of
pine pollen reaches a peak of 50% before it gradually
begins to decrease in the upper deposits of Period II.
The high percentages of pine pollen during these two
alluvial periods probably came from scattered pine trees
in Mexico and the Edwards and Stockton plateaus (Figure 1).
Some of the pine grains may have been carried to the site
as part of the Rio Grande alluvial deposits but most of
them were probably transported to the site by the wind.
This is the probable source of Pinus pollen deposition for
several reasons. In flood waters conifer pollen tends to
float on the surface due to their buoyant structure and
therefore tends to be deposited either in the delta at the

end of a river or in shore drift deposits. In either
case most of the pollen would be oxidized. In studies
conducted in the Whitewater Draw region, samples taken
from fresh alluvial deposits contained only a small
fraction of the conifer pollen found in surface froth
collected at the same time in nearby tributaries (Martin,
1963). Another reason why most of the pine pollen deposited during this alluvial period probably came from
windblown sources is evident in the pollen diagrams
from nearby rockshelters. During the corresponding time
period high percentages of pine pollen are also found
in the deposits from Bonfire Shelter, Eagle Cave, and
Centipede Cave (Johnson, 1963; Hevly, herein; McAndrews
and Larson, herein).

During the deposition of the lower alluvial deposits
of time Period II the boundary of pine trees was probably
much closer to the Devil's Mouth Site than it is today.
As the climatic conditions slowly began to change towards
the end of Period II, the boundary of pine trees may have
started to recede away from the site. Furthermore, any
climatic change in the area could have upset the delicate
balance of nature and thus prevented pine seedlings from
growing in the protected canyons along the rivers in the
Amistad region. Eventually the adult pines in these canyons died thereby eliminating another source of pine
pollen. Any, or all, of these factors could have been
responsible for the gradual decrease in Pinus pollen
during the latter part of Period II.

The high percentages of Nyctaginaceae in the pollen
record of Periods I and II suggests that the Amistad
region was subject to periodic heavy spring and summer
rains during part of the Altithermal Period. Alluvial
soils and heavy spring rains would tend to favor a large
Nytaginaceae plant population on terraces near the Rio
Grande. Further, the Karst Topography in the upland
areas would have caused the heavy summer rains to flow
quickly into the rivers, causing them to flood the nearby
terraces. This periodic flooding could have covered
large numbers of blooming Nyctaginaceae plants with
alluvium. If this occurred, then it would be recorded in
the pollen record as an overrepresentation of Nyctaginaceae pollen since many of the flowers would have been
buried under alluvium before their pollen could
be dispersed.

The low percentages of xerophytic plant pollen in
the fossil record of Periods I and II adds further support to the theory of a mesic Altithermal (Martin, 1963).

None of the deposits from these two periods contained
Prosopis, Acacia, Mimosa, Agave, Flouquieria, or Yucca
pollen, yet all of these plants currently exist in the
Amistad region and pollen of most of these genera is
found in the soil samples of upper Period III through
Period VII deposits. The only xerophytic plant represented in the pollen record of Periods I and II is Opuntia.
Cactus pollen is absent from Period I but is weakly represented in the lower deposits of Period II. Evidently
as the climatic conditions became less mesic certain
xerophytic plants began to invade the Amistad region.
The influx of certain types of cactus pollen into the
fossil record during this period coincides with the
gradual decrease of both pine and Nyctaginaceae pollen.

At least two other pollen types found in the
sediments of Periods I and II deserve mentioning, Liquidambar and Cucurbita. The presence of Liquidambar pollen in
Altithermal deposits at the Devil's Mouth Site remains a
mystery. Under phase-contrast analysis the fossil
Liquidambar pollen grain from Stratum 19 compared favorably with modern Liquidambar grains in our Amistad Reference
Collection. Furthermore, during conversations with Richard
Hevly, he mentioned that he had found several badly
deteriorated fossil grains in the samples from Bonfire
Shelter which could be Liquidambar pollen. If this is
true then perhaps Blair is correct in stating that sweetgum trees were probably living in this area of Texas in
prehistoric times (Blair, 1958). However, one or a few,
fossil Liquidambar pollen grains does not prove or dis-
prove the existence of sweetgum trees in the Amistad area
since the presence of a single pollen grain could be due
to long-distance transport or redeposition. Nevertheless,
the presence of Liquidambar in the fossil record does open
the door for further investigation into the subject.

The occurrence of Cucurbita pollen in Strata 18 and
19 is easier to explain than the occurrence of sweetgum
pollen. The fossil pollen record from Bonfire Shelter
notes the presence of Cucurbita pollen in several strata
dating prior to 7,000 B.C. The gourd pollen from the
Devil's Mouth Site does not date that early in time, but
it does indicate that gourds have been growing in the
Amistad Reservoir area for thousands of years (Whitaker,
1965).

During Period III (7,000-4,000 B.C.) there was an
indeterminate period of erosion which severely eroded the
surface and edges of the river terraces in the Amistad
region. Unfortunately, the stratigraphic evidence at the

Devil's Mouth Site did not reveal the time span of this erosional period. However, the pollen record does indicate that there were some marked vegetational changes during this period. These changes can be seen in an examination of the pollen record of the pre-erosional and post-erosional deposits at the Devil's Mouth Site and in the fossil pollen diagrams from rockshelters in the Amistad region. The pollen diagram from the Devil's Mouth Site indicates a slight rise in the percentage of Gramineae and Ephedra pollen immediately prior to the erosional period and a sharp decrease in both types after the erosional period. The pollen analysis of Bonfire Shelter revealed a similar increase and decrease of these types during approximately the same period (Hevly, herein). The analysis of Centipede Cave shows an increase and decrease of Gramineae during this erosional period but the percentages of Ephedra pollen remained fairly constant (Johnson, 1963). The pollen record for this time period is incomplete from Eagle Cave. Thus the correlative evidence, although not conclusive, indicates that this severe erosional period occurred during a xeric climatic interval (evidenced by an increase in the percentages of Ephedra and grass pollen).

Following the erosional period the Amistad region became gradually more xeric. The pollen from certain plant types such as Prosopis, Mimosa, Acacia, and Agave first appear in the pollen record of post-erosional deposits. Later, during Periods IV through VII some of these pollen types such as Acacia and Mimosa show marked percentage increases illustrating that the climatic conditions in the region were probably most favorable for the growth of semi-arid plants. This assumption is also supported by the pollen record of Ephedra and Opuntia. During these time periods both pollen types became important contributors to the fossil pollen record of the Devil's Mouth Site whereas during the prior mesic interval of Periods I and II both types were weakly represented.

One of the significant differences between the pre-erosional and post-erosional pollen record is the pollen curve of the Cheno-Ams. In the pre-erosional deposits the percentages of Cheno-Am pollen do not reveal any surprising changes in the secondary pollen count. However, the increases in the percentages of Cheno-Am pollen in the post-erosional deposits indicate that Cheno-Ams were most prevalent during Periods III through VII. Plants classified by their pollen morphology as Cheno-Ams thrive

in alkali soils and grow well in midden areas where the
soil has been disturbed (Iversen, 1941). At the Devil's
Mouth Site every significant increase in the percentages
of Cheno-Am pollen in the fossil record occurred either
during or immediately after known periods of intense occupation by primitive man (Johnson, 1963).

For a short time during Period V (1,000-200 B.C.)
the climate in the Amistad region may have approached
mesic conditions. The pollen record from Strata 8 and 9
indicates a sharp rise in pine, Nyctaginaceae, and Cyperaceae pollen and a general decrease among the Cheno-Ams,
Ephedra, and Prosopis. The return to mesic conditions,
if this interpretation of the data is correct, was short-
lived for subsequent pollen deposition indicates a return
to xeric conditions. Thus the xeric conditions of the
post-erosional period lasted for approximately 4,500
years and was interrupted only briefly during the deposition of Strata 8 and 9.

The modern surface sample collected on the terrace
contained high levels of both arboreal and anemophilous
pollen. The pine pollen consisted almost entirely of
small grains which compared morphologically with reference
pollen grains of the Pinus edulis variety. The pine pollen was transported to the site from any or all of the
following possible sources: 1) unknown locations in nearby Mexico, 2) pine trees currently growing in a few protected canyons in Mexico and the Transpecos region of
Texas, and 3) isolated pines growing on the higher eleva-
tions of the Edwards and Stockton plateaus. Johnson
states that he found pines growing 40 miles north of the
Amistad Reservoir (Johnson, 1963; see also report herein
by Flyr). However, the nearest pine trees the writer was
able to locate during the preliminary study of the flora
in the Amistad region, were approximately 75-100 miles
north of the area. In terms of the modern pollen rains it
seems likely that most of the pine pollen is coming from
sources in Mexico since during the pine pollination period
the prevailing surface winds in the Amistad region average
11.6 miles per hour and blow predominately from the
southeast (U.S. Dept. of Commerce, 1964). Juniperus pollen
is also transported a long distance to the site since there
are no cedar trees growing on or near the site. Part of
the cedar pollen is probably coming from unknown sources
in Mexico while the remainder is coming from trees located
approximately 35 miles northwest of the site on the Edwards
and Stockton plateaus. The Juglans, Carya, and Quercus pollen is probably coming from sources along the eastern shore

of the Devils River where these trees can be seen growing in scattered groups along the terrace. The small percentages of other arboreal pollen types such as Prosopis, Mimosa, and Acacia, come from trees growing on the terrace near the Devil's Mouth Site. However, it is surprising that these Legumes are so weakly represented in the modern sample since they grow in great abundance along the full length of the terrace.

Compositae and Opuntia pollen are weakly represented in the terrace surface sample. In the sample from Stratum 1, 82% of the pollen came from either Opuntia or Compositae sources. However, the surface sample shows a sharp reduction in both types. Compositae pollen drops from 47% to 27% while Opuntia pollen drops from 35% to 2%. The decrease in Compositae pollen, as compared to Stratum 1, can be attributed to recent overgrazing of livestock in the area. The low percentage of Opuntia is possibly the normal representation of a surface sample from the terrace. Large patches of Opuntia leptocaulis and Opuntia lindheimeri currently grow on the terrace and were probably present in approximately the same amounts during the deposition of Stratum 1. The high levels of Opuntia pollen in the deposits of Stratum 1 may indicate an intraflorae caused either by nature or man.

The surface sample from the terrace also contains small percentages of Nyctaginaceae, Cheno-Ams, Malvaceae, and Ephedra pollen. All of these plant types, except Ephedra, are currently growing either on the terrace or in the eroded gullies along the terrace edge. Ephedra grows in the upland regions a short distance from the site.

The other two surface samples, one taken from the ridge northwest of the site and the other taken from the stock tank on the Figueroa ranch, tended to support the pollen diagram of the modern terrace sample (Figure 23). The sample collected from the rim of the limestone ridge contained 67% arboreal pollen, all of which was transported there either by the wind or by insect sources since none of the plant types grow along the ridge surface. One surprising aspect of the rim sample was the absence of Agave, Yucca, and Fouquieria since these are the dominant plant types currently growing on the rim surface. The stock tank sample, like the other surface samples, contained mainly anemorphilous pollen from trees and shrubs.

The high levels of arboreal pollen in the surface
samples do not accurately represent the present types of
trees and shrubs growing in the vicinity of the Devil's
Mouth Site. Instead, the pollen diagram of the modern
soil samples results from the near absence of surface
vegetation. For example, 42% of the pollen in the surface
sample from the terrace comes from plants which do not
grow on or near the terrace. Plant denudation is often
the result of cattle, goat, and sheep grazing. In climatic fringe areas with low rainfall and high temperatures
(for instance, the Amistad region), overgrazing is much
more damaging to plant cover than in more mesic areas
where weedy plants quickly replace natural vegetation.
For years ranchers in the Amistad region have allowed
livestock to browse for the few edible plants in the area.
The result of this is clearly evident in both the present
vegetation and the pollen record. The vegetation in this
region is presently characterized mostly by drought and
animal resistant plants such as Yucca, Larrea, Opuntia,
Fouquieria, Mammillaria, Acacia, Prosopis, Mimosa, and
Agave. The modern pollen record also shows a general
reduction in the pollen percentages of plants which are
normally considered edible by livestock. A more nearly
accurate record of what the vegetation was probably like
prior to the introduction of ranching is reflected by the
pollen record of Stratum 1 since it contains higher percentages of pollen from plants which are normally eaten
by livestock.

       *   *   *   *   *   *   *   *   *   *

One of the most difficult problems in any palynological study of an archeological site is the removal of
microscopic particles of charcoal from the pollen samples.
Some of the pollen samples from the Devil's Mouth Site
contained large amounts of charcoal. This caused most
standard charcoal-removing techniques to be ineffective
on the pollen samples and necessitated the development of
a new charcoal-removing technique.

Some palynologists recommend that soil samples from
archeological sites be finely crushed in a mortar and
pestle and then screened through a fine mesh wire screen.
This technique is supposed to remove much of the charcoal,
and, in fact, does. However, it also selectively destroys
certain pollen types (Picea, Pinus, Opuntia). The fossil
pollen from the Devil's Mouth Site was very poorly preserved and after using this technique many of the pollen
grains over forty microns were broken or crushed.

Another standard carbon-removing technique is to
heat the sample in a solution of 10% nitric acid. This
method proved useful, but did not remove sufficient
amounts of charcoal from the samples. Repeated nitric
acid treatments removed additional quantities of charcoal,
but it also deteriorated the surface structuring of the
grains, especially the echinate grains. After repeated
treatments with 10% nitric acid the writer found that
many of the echinate grains were so badly deteriorated
that accurate identification was impossible.

Through experimentation an improved charcoal-removal
technique was developed which proved useful in the processing of the samples from the Devil's Mouth Site. A
heated solution of 10% potassium hydroxide used before
hydrofluoric acid removed large amounts of charcoal.
Several of these potassium hydroxide treatments generally
removed sufficient quantities of charcoal, making possible
an accurate microscopic analysis of each sample. Experimentation showed that the use of potassium hydroxide after
hydrofluoric acid did not have the same effect upon the
charcoal. Perhaps some of the silicates which dissolve
during the hydrofluoric treatment impregnate the charcoal
particles, thereby impeding proper reaction with the
potassium hydroxide.

SUMMARY

The pollen analysis of the Devil's Mouth Site was
conducted in an effort to determine the late Quaternary
climate in the Amistad region. A total of twenty-eight
fossil soil samples were processed, using a newly-developed
digestion technique in order to recover sufficient pollen
from each sample for an analysis. Preliminary analysis
of the fossil samples revealed a dominance of a single
pollen type, Compositae, throughout the late Quaternary
period of the site. A second count, excluding Compositae
pollen, was done in order to ascertain a better understanding of the prehistoric vegetation.

The pollen evidence from the Devil's Mouth Site
revealed several climatic changes in the Amistad Reservoir
area. During time Periods I, II, and part of III (7,000-
3,000 B.C.), the region had a mesic environment which was
probably caused by increased amounts of spring and summer
rainfall. This mesic interval was terminated during late
Period III (ca. 3,000-2,500 B.C.) by a marked period of
erosion which severely eroded the river terraces along
the Rio Grande.

Following the erosional period the climate in the Amistad region changed from mesic to xeric. This is seen in the pollen record as increases in the percentages of pollen from xerophytic plants. Xeric conditions prevailed in the area until midway through time Period V, when the climate became slightly mesic. However, this short mesic interval was soon terminated and the climate again became xeric.

APPENDIX A

PRELIMINARY STUDY OF THE LATE QUATERNARY CLIMATES IN

THE AMISTAD RESERVOIR AREA OF SOUTHWEST TEXAS,

THE POLLEN EVIDENCE

The fossil pollen records from two rockshelters (Bonfire Shelter and Eagle Cave) and one terrace site (Devil's Mouth Site) contain many similarities. Through careful examination of manuscripts in this report and other miscellaneous notes and pollen diagrams not incorporated into this report, it is possible to reconstruct a tentative climatic sequence for the late Quaternary period in the Amistad Reservoir area of southwest Texas.

In the following discussion of climatic conditions in the Amistad region during the last 10,000-12,000 years, the pollen records of these three aforementioned sites are examined in terms of the eight major time periods for the region outlined by Dr. Dee Ann Story in this report.

<u>Periods I and II</u> (prior to 7,000 to 4,000 B.C.--Mesic conditions prevail.)

The pollen diagrams from Bonfire Shelter, Eagle Cave, and the Devil's Mouth Site show evidence of a pine peak just prior to the termination of time Period I and a second pine peak immediately following the beginning of Period II. (At Eagle Cave the second pine peak occurs in the undated deposits of the upper Sterile Stratum. Being void of cultural remains this stratum was not assigned by Story to a time period; however, the pollen evidence suggests that at least the upper portion of the Sterile Stratum should be tentatively assigned to early Period II.) During the remainder of Period II, the percentage of pine pollen gradually decreases in the pollen record of each of these three sites. Other noticeable areas of palynological comparison among these three sites during Periods I and II include Cheno-Ams, <u>Ephedra</u>, and Onagraceae. On the pollen diagrams from each of these sites <u>Ephedra</u> and Cheno-Am pollen grains are weakly represented during Period I but increase during Period II; the opposite is true of the Onagraceae pollen record during these two time periods.

154

At Bonfire Shelter and the Devil's Mouth Site there
is a grass pollen peak in the deposits immediately
preceding the close of Period I, a slow and gradual
increase in grass pollen during Period II, and a
second grass pollen peak following the close of
Period II. The pollen record from Eagle Cave reveals
a similar grass pollen peak prior to the termination
of Period I but shows a second peak early in Period
II. It is interesting to note that all three pollen
records, as tentatively correlated by our laboratory,
agree in one respect: each of them indicates a noticeable percentage increase in grass pollen prior
to and during the deposition of the bison bones in
Bone Bed 2 of Bonfire Shelter.

Periods III and IV   (4,000 to 1,000 B.C.--Period of
                       climatic change in the Amistad
                       region. The mesic conditions
                       of the previous periods terminate
                       and are replaced by a xeric
                       environment).

The vegetational and climatic history of
these two time periods is difficult to interpret
since none of the sites offers an adequate or complete pollen record of these periods. Part of the
sediments deposited during Period III are not represented at the Devil's Mouth Site. Only a partial
pollen record of this period is available from Bonfire Shelter and Eagle Cave since some of the sediments did not contain sufficient pollen for analysis.
At the Devil's Mouth Site the pollen record of
early Period III deposits reflects vegetational
conditions which could be classified as marginally
mesic. Late Period III and Period IV deposits from
the same site contain a pollen spectra characteristic of a xeric environment. Though incomplete, the
pollen records that are available for Periods III
and IV from Eagle Cave and Bonfire Shelter reveal an
increase in Ephedra, Prosopis, grass, Compositae,
Cheno-Ams, and Agave pollen and a definite decrease
in pine and Nyctaginaceae pollen. In comparison, the
pollen record from Devil's Mouth Site also indicates similar fluctuations in the aforementioned
pollen types during these time periods with definite
evidence of decreases in pine and Nyctaginaceae pollen.

Period V   (1,000 to 200 B.C.--Xeric conditions prevail.)

The Devil's Mouth Site deposits offer the only
complete record of this period. In corresponding deposits from Eagle Cave pollen is absent probably
because it was destroyed by the fires of primitive

man. Pollen data for Period V deposits from Bonfire
Shelter are incomplete since some of the samples are
void of pollen. The pollen record from the Devil's
Mouth Site indicates that during early Period V
there is a noticeable increase in pine, Nyctaginaceae, and sedge pollen. The equivalent portion of
Period V in Bonfire Shelter shows an increase in
pine, sedge, Prosopis, riparian trees, and grass pollen and a general decrease in Ephedra and Cheno-Am
pollen. This short mesic interval corresponds to
Bone Bed 3 in Bonfire Shelter and may have afforded
climatic and vegetational conditions in the Amistad
region suitable for large bison herds. During the
remainder of Period V the pollen records of Bonfire
Shelter and the Devil's Mouth Site indicate that
the climate in the Amistad region returned to
xeric conditions.

Periods VI and VII (200 B.C. to A.D. 1,600--Continued
xeric conditions.)

At the Devil's Mouth Site the pollen record of
Periods VI and VII are characterized by decreasing
percentages of pine, Nyctaginaceae, Cyperaceae, and
Ephedra pollen, along with noticeable increases in
cedar, mesquite, Acacia/Mimosa, Cheno-Am, and cactus
pollen. At Bonfire Shelter the deposits of Zone 3b
(above the Fiber Layer) have tentatively been assigned to Period VI. The pollen evidence from these
deposits at Bonfire Shelter reflects a record of
xeric conditions characterized by decreases in pine
and Ephedra and increases in cedar, oak, mesquite,
grass, Cheno-Am, and Agave pollen.

Historic Period, Period VIII (A.D. 1,600 to present--
Xeric conditions prevail.)

The pollen record from terraces in the Amistad
region offer the most complete record of recent
vegetational changes caused by man. The pollen
spectra from Stratum 1 and the surface of the terrace at the Devil's Mouth Site show high percentages
of pollen from drought and animal resistant plants
such as cactus, mesquite, and Acacia/Mimosa and low
percentages of pollen from plants (Cheno-Ams, Compositae, and grass) which are considered edible or
subject to mechanical damage by livestock. Another
indication of recent plant denudation by livestock
in the Amistad Reservoir area is reflected by the
high percentages of arboreal anemophilous pollen of
long-distance transport (pine and cedar) in contemporary pollen samples.

APPENDIX B

All of the pollen samples from the Devil's Mouth Site were processed in the Botany Department's Palynological Laboratory located on the campus of The University of Texas. Extraction procedures were conducted in a fume chamber in order to prevent harmful damage caused by the inhalation of certain acid fumes.

EXTRACTION TECHNIQUES

I.   Removal of large particles of organic and mineral origin.

   A.   Remove approximately five ounces of sediment from each sample and screen through a one millimeter mesh screen. This removes small rocks and plant particles from the sample.

II. Removal of calcium and carbonates from the samples.

   A.   Place the screened sample in a 600-800 milliliter beaker and fill one-third full with distilled water. Then mix the solution thoroughly. When this is completed add concentrated hydrochloric (HCl) to the solution until all reaction ceases. If the reaction becomes violent, add either acetone or 95% ethnol to the solution in order to break the surface tension of the liquid.

III. Removal of coarse-grained silicates.

   A.   Corase-grained silicates are removed by decanting.

      1.   Stir the solution rapidly in all directions in order to prevent the formation of water currents. Allow the sample to stand for 30 seconds before decanting. In decanting, pour the aqueous fraction into a separate container and discard the coarse sediments in the bottom of the beaker.

157

2. Resuspend and then decant the solution a second time in the same manner.

3. Resuspend and decant the aqueous fraction a third time in like manner, only allow the mixture to stand for one and one-half minutes before decanting.

4. After the third decanting, place the aqueous fraction into a plastic acid-resistant 100-milliliter centrifuge tube and centrifuge at 2,500 RPM for one minute.

5. Pour out aqueous fraction and save solid residue.

IV. Removal of charcoal.

A. Once the corase-grained silicates are removed, then treat the sample with 10% potassium hydroxide (KOH) to remove the charcoal and carbon particles from the sample. This procedure will also soften other cellulose particles so that they can be removed more easily by acetolysis.

1. Mix the sample with 500 milliliters of distilled water and then screen through a brass 200-micron mesh screen in order to remove the larger particles of charcoal.

2. Centrifuge and then discard aqueous fraction.

3. Wash with distilled water, centrifuge, and discard liquid fraction.

4. Repeat water wash procedure several times.

5. After washing is complete, fill the centrifuge tube containing the sample with a solution of 10% KOH and place in a boiling water bath for five minutes.

6. Remove the sample from the waterbath and centrifuge. After centrifuging, discard liquid fraction.

7. Wash with distilled water, stir, centrifuge, and decant aqueous fraction.

8. Repeat wash procedure five to ten times or until liquid fraction remains clear after centrifuging.

V. Solution of fine-grained silicates and removal of colloids.

A. Removal of silicates.

1. Place sample in hydrofluoric (HF) resistant centrifuge tube. Add five milliliters of HF and stir vigorously for ten seconds. Occasionally, mica and other particles which are not removed during the HCl treatment have a violent reaction when they come in contact with HF. This reaction does not occur at room temperature. Instead, the reaction occurs when the HF reaches a high temperature. In most samples this violent reaction occurs approximately 45 seconds after the addition of HF. However, in no instance has it occurred in less than 25 seconds after the addition of HF. Occasionally, it is necessary to add small amounts of 95% ethnol in order to reduce excessive boiling. This will prevent the sample from excaping the confines of the centrifuge tube. Once the initial reaction ceases, continue to add small amounts of HF until the centrifuge tube is approximately one-third full.

2. Place the sample containing the HF into a boiling waterbath for 20 minutes.

3. Remove from waterbath and centrifuge. Save solid fraction and discard liquid fraction.

B. Removal of colloids formed during
HF procedure.

1. Fill the centrifuge tube containing
the sample with concentrated HCl,
then mix thoroughly. Place sample
in boiling water bath until HCl begins to boil (approximately 60-90
seconds).

2. Remove from waterbath and centrifuge.
Decant aqueous fraction.

3. If any colloids remains, then wash
the sample in a solution of 10% HCl.
Centrifuge and discard liquid fraction. Repeat this washing procedure
until all colloids are removed (approximately 1-7 washed depending
upon the size and type of sample).

VI. Removal of the organic fraction.

A. Acetolysis.

1. After removing the colloids, wash
the sample in glacial acetic acid to
remove the water.

2. Centrifuge and decant aqueous fraction.

3. Add 30 milliliters of a mixture of
nine parts acetic anydride to one part
concentrated sulfuric acid to the
sample. Stir sample thoroughly and
then place in a boiling water bath for
five minutes. Take care not to let
water from the water bath get into the
acetolysis mixture.

4. Remove and centrifuge. Decant aqueous fraction.

5. Fill centrifuge tube with glacial acetic
acid and stir. Centrifuge and decant
liquid fraction.

6. Wash sample four times with distilled
water. After each wash decant and
discard aqueous fraction.

VII. Additional procedures.

    A. When the standard procedure outlined above has been completed, then examine each sample under a microscope to see what additional steps might be needed before the material is ready for analysis. If, for example, the sample still contains small particles of silicates, then repeat the HF treatment. If cellulose particles remain, then acetolyze the sample. However, if charcoal remains, do not repeat the KOH procedure. The use of KOH after acetolysis sometimes damages the pollen grains.

APPENDIX C

SLIDE PREPARATION

I.   Staining of palynomorphs.

    A.   Safranin staining.

        1.   Place sample in a 12-milliliter glass
             centrifuge tube.

        2.   Fill one-half full with 95% ethnol.

        3.   Add 5 drops of .2% safranin stain (in
             a medium of 70% ethnol) and stir
             thoroughly.

        4.   Allow the sample to remain in the
             stain for two minutes.

        5.   Centrifuge and decant liquid fraction.

        6.   Wash in 95% ethnol. Centrifuge and
             decant aqueous fraction.

II. Mounting and mounting media.

    A.   Silicone oil media.

        1.   Suspend sample in 95% ethnol. Centri-
             fuge and decant liquid fraction.

        2.   Suspend sample in 100% ethnol. Centri-
             fuge and decant liquid fraction.

        3.   Transfer the sample to a five milliliter
             shell vial and fill three-fourths full
             with benzene. Centrifuge and decant
             benzene.

        4.   Add a few drops of 2000 cs silicone
             oil to the shell vial and stir until
             the silicone oil is thoroughly mixed
             with the sample.

162

5. Mark the shell vial with a diamond pencil and label the cork with a pen. (Both cork and vial are marked in order to insure accuracy and to prevent any chance of mistaken identity.)

6. Place the shell vial containing the silicone oil on a warming plate for several hours in order to allow the remaining benzene to evaporate. When this is completed the sample is ready to be mounted on a microscope slide.

B. Mounting on slides.

1. Place on drop of silicone oil from the vial on a clean glass microscope slide.

2. Clean a No. 1. 22mm^2 glass cover slip and place it carefully on top of the drop of silicone oil.

3. When the silicone oil spreads to the edges of the cover slip, then seal the cover slip with nail polish.

4. Label each slide and store for future analysis.

A PRELIMINARY POLLEN ANALYSIS OF BONFIRE SHELTER

Richard H. Hevly

Excavations at Bonfire Shelter in 1963-64 revealed a stratified bison kill site containing evidence of intermittent human use for at least the last 10,000 years, but at the same time raised questions as to the nature of the environment supporting the prehistoric aborigines and the herds of large herbivores hunted by them in this arid region presently dominated by Chihuahuan Desert scrub vegetation (Blair, 1950; Dibble, 1965). In an attempt to answer this and other questions resulting from archeological studies in the Amistad Reservoir, palynological studies of the shelter sediments were inaugurated as previous studies of Southwestern cave sediments had demonstrated the potential of utilizing fossil pollen in the reconstruction of late Quaternary vegetation and climates (Anderson, 1955; Johnson, 1963; Laudermilk and Munz, 1934; Martin, Sabels, and Shutler, 1961; Sears and Roosma, 1961).

Furthermore, analysis of trial samples submitted to me at the Geochronology Laboratories of The University of Arizona in 1964 indicated that the sediments of this shelter contained a pollen record suitable not only for the elucidation of the environmental potential afforded the prehistoric aborigines and their associated fauna, but also the opportunity to examine the history of utilization and perhaps domestication of native plants (i.e., Cucurbita) by the prehistoric inhabitants of this rockshelter.

MATERIALS AND METHODS

In early June, 1965, sediment samples for pollen analysis were obtained from the vertical, cleaned walls of two excavation areas (pits) with the assistance of David S. Dibble and Vaughn M. Bryant, Jr., utilizing techniques described by Schoenwetter (1960) for arroyo walls. Modern pollen samples for comparison with the fossil record were also collected from cattle tanks, water-filled potholes in canyon bottoms, and ponds, as well as soil surfaces of five different plant associations in the Amistad Reservoir and Edwards Plateau areas, following standard methods (Bent and Wright, 1963; Hafsten, 1961; Hevly and Martin, 1961; Hevly, Mehringer, and Yocum, 1965).

The sediment samples were extracted in the Palynological Laboratory of the Department of Botany, The University of Texas, by Vaughn Bryant, following standard techniques (Faegri and Iverson, 1964). The resulting pollen concentrate was mounted in glycerine and analyzed microscopically at a magnification of 400 or 1000 diameters. Standard keys and illustrations provided by Faegri and Iverson (1964) and Erdtman (1957) were utilized for pollen identification and augmented by comparison of fossil pollen types with a pollen reference collection of extant species occurring in the Amistad Reservoir area prepared by Vaughn Bryant and John H. McAndrews.

RESULTS

A. Modern Pollen Studies - All but three of the fourteen modern pollen samples yielded sufficient pollen for a basic 200-grain count from which aquatic plant pollen and all spores were excluded (73% yield). Fifty-four pollen types were identified, and, although no well-preserved grains remained unknown, up to 10% of the pollen in several samples was too poorly preserved for identification (Table 7). In sample #8, riparian trees (principally Carya) constituted about 76% of the pollen so this sample was reanalyzed excluding a riparian tree category (including Alnus, Betula, Celtis, Carya, Fraxinus, Juglans, Platanus, Populus, Sapindus, and Salix).

The Transpecos-Shrub Savanna and Ceniza Shrub plant associations of the Amistad Reservoir are generally characterized by low relative abundance of arboreal pollen (less than 35%), while the Mesquite Savanna, Mesquite Oak or Juniper Oak Savanna and Oak Juniper Woodland are characterized by more than 35% arboreal pollen (Figure 24). The Mesquite Savanna samples of this study yielded about 40% arboreal pollen, while the other higher elevation plant associations yielded 55% or more arboreal pollen. Throughout the modern pollen samples Prosopis and riparian tree pollen percentages remained relatively uniform; however, Juniperus and Quercus pollen percentages increased with elevation, and Pinus pollen exhibited maxima at both extremes of the plant associations studied. Pine pollen exceeded 20% only near or where pine trees grew and where vegetation was sparse or wind-pollinated types essentially absent. In the latter samples which came from tanks or ponds in the Transpecos-Shrub Savanna plant association, pollen of a variety of desert plants is also present, thus providing a means of distinguishing the high pine records of xeric habitats from the more mesic habitats where desert plants are generally absent.

TABLE 7. MODERN POLLEN SAMPLES

Sample Designation	Type	Vegetation Description	Location
VB	Transpecos Shrub-Savanna (?)	- - -	stock tank on Figueroa Ranch
32	Transpecos Shrub-Savanna (?)	- - -	beaver pond near Devils River
3	Transpecos Shrub-Savanna	cresote bush + riparian trees	pothole in Mile Canyon, 1/4 mile from Rio Grande
1	Transpecos Shrub-Savanna	cresote bush, mesquite, tarbush, ocotillo, agave, cholla, yucca, acacia	notch above Bonfire Shelter, rim of Mile Canyon
4	Transpecos Shrub-Savanna	cresote bush, mesquite, tarbush, ocotillo, agave, cholla, yucca, acacia	Hwy. 90, 5 miles east of Langtry, Texas
2 & 6	Ceniza Shrub	cresote bush, ceniza and mesquite + sotol, nolina and barberry	seven miles from Comstock, Texas, on Pandale road
12	Mesquite-Savanna	including juniper and barberry	eighteen miles from Comstock, Texas, on Pandale road
7	Mesquite-Savanna	including juniper + barberry	twenty-three miles from Comstock, Texas, on Pandale road
5	Oak-Juniper Woodland	including pinyon pine + barberry	twenty miles north of Brackettville, Texas
8	Mesquite-Oak Savanna	+crosstimbers riparian-- juniper-oak savanna ecotone + barberry	along Llano River, 5 miles south of Junction, Texas

TABLE 7 (cont'd)

Sample Designation	Type	Vegetation Description	Location
9*	Transpecos Shrub-Savanna	- - -	pool at head of Mile Canyon
10*	Transpecos Shrub-Savanna	- - -	mouth of Mile Canyon
11*	- - -	- - -	limestone from wall of Bonfire Shelter

*No pollen recovered from sample

The non-arboreal record also exhibited useful characteristics for distinguishing the different plant associations. Gramineae pollen, for example, parallels the Juniperus record achieving greatest relative abundance in the high elevation savanna and woodland communities and exceeding 12% only in those communities in which oak is an important component. As the relative abundance of Gramineae pollen declines with elevation, the percentages of Ephedra and other desert shrubs such as Acacia, Mimosa, Larrea, Fouquieria, Koeberlinia, and Leucophyllum increase, constituting 7-20% of samples from the Transpecos-Shrub Savanna and Ceniza Shrub plant associations but 5% or less of the Mesquite Savanna, Mesquite Oak or Juniper Oak Savanna and Juniper Oak Woodland. With the exception of Agave, Larrea, Fouqueria, Jatropha, Koeberlinia, Leucophyllum, and Opuntia, which appear only in samples from the Transpecos-Shrub Savanna and Ceniza Shrub plant associations, pollen of other desert plants such as Sphaeralcea, Boerhaavia, or Mirabilis, Gaura, or Oenothera and Yucca, Dasylirion, and Nolina is never abundant and frequently sporadic in occurrence. Ephedra occurred in nearly all modern pollen samples and was always the torreyana type; however, this was not unexpected as only species producing this type were encountered while collecting modern pollen samples, and Martin (1963) has noted that species producing the nevadensis type of Ephedra pollen are restricted to the western desert areas of North America where a winter rainfall pattern prevails.

The summer dominant rainfall pattern characteristic of the Amistad Reservoir may also be indicated by another non-arboreal pollen type, the Cheno-Ams, whose relative abundance achieved only 3-15%, not assuming the high abundance characteristic of cattle tank and many soil surface samples elsewhere in the Southwest. This phenomena was not surprising as Cheno-Ams are generally most abundant in alkaline soils of winter-moist desert areas and a scarcity of species belonging to this group was noted while collecting modern pollen samples.

Instead of domination by Cheno-Ams, the non-arboreal pollen of the Amistad Reservoir was composed predominantly of Compositae which accounted for 20-40% of the grains observed. The Compositae were separated into the usual high and low spine categories after removal of Liguliflorae and Artemisia, pollen of the latter genus being recovered only in Mesquite Savanna, Mesquite Oak or Juniper Oak Savanna and Juniper Oak Woodland (Hevly, Mehringer, and Yocum, 1965).

FIGURE 24. Modern pollen rain of the Amistad Reservoir and Edwards Plateau.

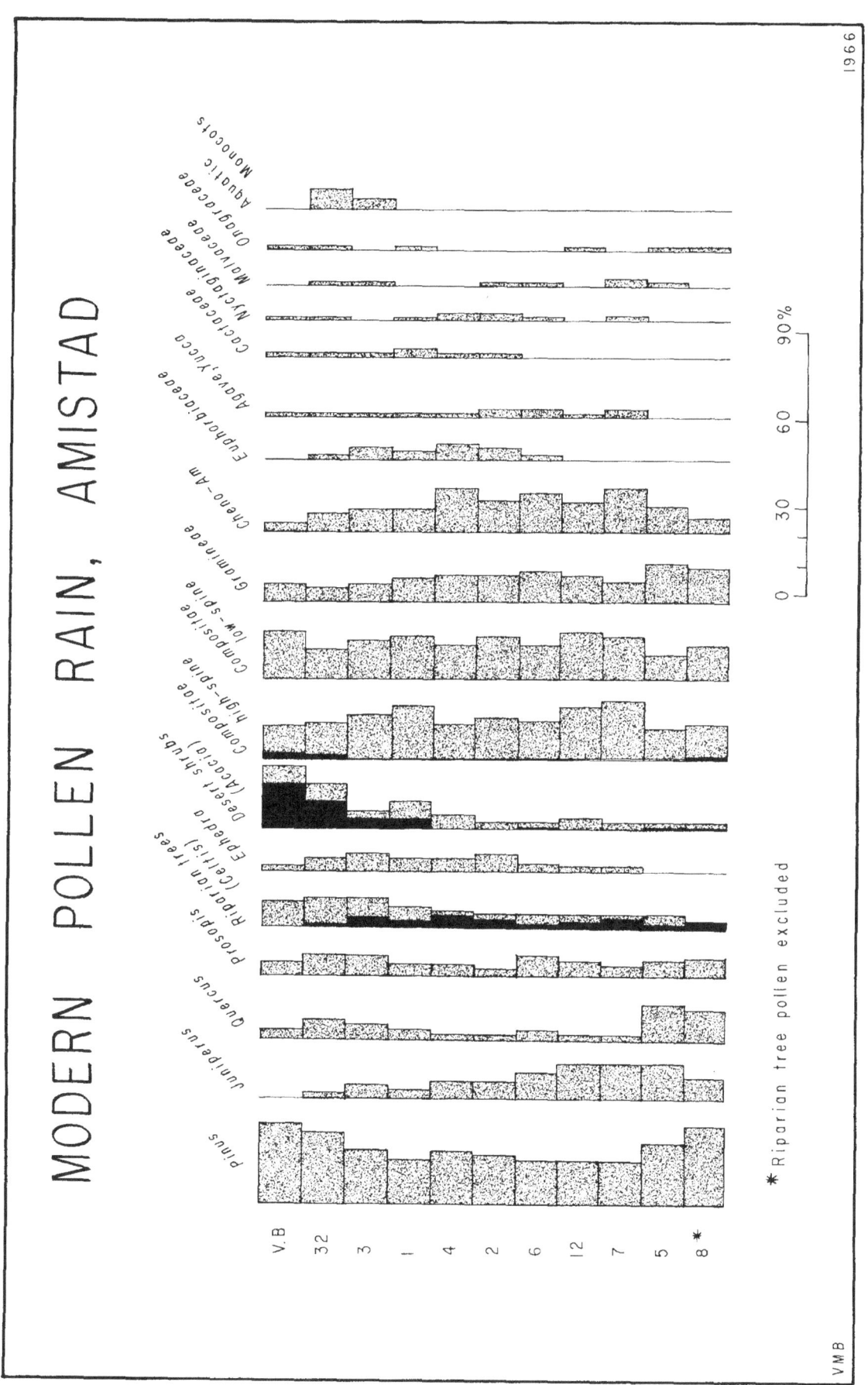

MODERN POLLEN RAIN, AMISTAD

* Riparian tree pollen excluded

1966

VMB

B. _Fossil Pollen Studies_ - Bonfire Shelter
yielded incomplete but overlapping and complementing
fossil pollen records so that a general outline of
the history of the vegetation in the immediate area
during the past 10,000 years is now available. This
outline is based on three profiles, A, B, and C, collected from the following coordinates respectively:
N110-W42, N98-W47.5, and N30-W67 (Figure 16). The
location of these coordinates and the respective
stratigraphic units composing them are described by
Dibble (1965) in his report of the shelter.

Profile A, which was collected in 1964 by the
archeologists while excavating the site, consisted
of six samples for trial analysis from the major
stratigraphic units in descending order: Zone 3,
Bone Bed 3, Zone 2b, Bone Bed 2, Zone 1, Bone Bed 1.
Sufficient pollen for analysis was recovered from
Zones 3 and 2b as well as Bone Beds 1 and 2, but pollen was essentially absent in samples from Bone Bed 3
and Zone 1 (66.6% yield). Forty pollen types were
recovered and all of these, with the exception of
Tilia from Bone Bed 1, were also recovered in the
modern pollen rain; however, _Artemisia_ and Gramineae
in Bone Bed 1, _Ephedra_ in Zone 2 as well as Bone Bed 2,
and a number of large entomophilous pollen types, including Cucurbitaceae, Onagraceae, Cactaceae, Nyctaginaceae, Malvaceae, and Amaryllidaceae from the
cultural levels, particularly were more abundant than
in the modern pollen rain. _Ephedra_ pollen of the
nevadensis type was quite abundant in Bone Bed 2 but
is absent in the modern pollen rain. _Pinus_ pollen
was generally less abundant than in the modern pollen
rain but did exhibit an increase in relative abundance
in Bone Bed 1. Other pollen types occurred throughout
the fossil pollen record in about the same general
proportion as in the modern record.

Profile B, which was collected in 1965 from the
same general area of the site as Profile A, consisted
of 23 samples taken at variable but close intervals
through the 10.3 feet of sediment exposed in the wall
of a large, deep pit resulting from archeological excavation. In Profile B, adequately preserved pollen
was recovered in all but five samples that came from
Zone 2 (80.8% yield), and it is probable that further
experimentation with these recalcitrant samples will
eventually yield sufficient material for analysis.

Profile C, which was collected at the same time as Profile B, but which was located in the southern portion of the shelter (Figure 16), consisted of 12 samples collected from the 7.2 feet of wall remaining from excavation. Few samples were collected at this location due to the coarser texture of the sediments, extensive burning of the bone deposits, and heightened probability of redeposition of pollen from the nearby talus cone beneath a notch in the canyon rim overhanging the shelter. As expected, preservation was poor at this locality, but nine of the twelve samples did produce sufficient pollen for analysis (75% yield). At this locality, as at Profile A, Bone Bed 3 and Zone 1 did not produce adequate quantities of pollen nor did the lower portion of Bone Bed 2; however, while Zone 2 did not yield enough pollen for analysis at Profile B, it did so at Profile C, thus providing in Profiles B and C an overlapping pollen record affording some cross correlations from different areas of the shelter and at the same time augmenting one another to provide an essentially continuous record.

The pollen record from Bonfire Shelter, although incomplete, exhibits striking changes in the relative abundance of several pollen types, particularly _Pinus_, _Ephedra_, and certain entomophilous desert herbs. In fact, the two latter categories so far exceeded any modern records that their extremely high abundance has been regarded as abnormal and a probable artifact of human agency. For this reason, two extra pollen counts were prepared at Profile B excluding these types. The effect of the extraordinary abundance of these types seems primarily to have resulted in the suppression of the relative abundance of arboreal pollen and _Pinus_ in particular (Figure 25). An additional effect to compensate for this restraint was made by preparing separate 200-grain counts of arboreal and of non-arboreal pollen types (Figures 26 and 27). There emerges from this analysis a series of pollen zones which correspond closely with the easily discerned stratigraphy of the shelter sediments, and it is by these stratigraphic units or zones that the pollen results will be described.

Zone 1 is characterized by percentages of _Pinus_, _Juniperus_, _Quercus_, _Prosopis_, and Gramineae equal to or greater than the modern pollen rain of the immediate area, and in this zone a few sporadic grains of _Picea_ and _Pseudotsuga_ have also been recovered. Fossil Compositae pollen is about equal in abundance to that recovered from the modern pollen rain, but the fossil Cheno-Am pollen is remarkably less abundant than in the modern record. Other more

FIGURE 25. Fossil pollen record at Profile B (N=200).
            Note increase in pollen percentages when
            Ephedra and "economic" pollen types are
            removed from the standard 200-grain count.

BONFIRE SHELTER 41 VV 218

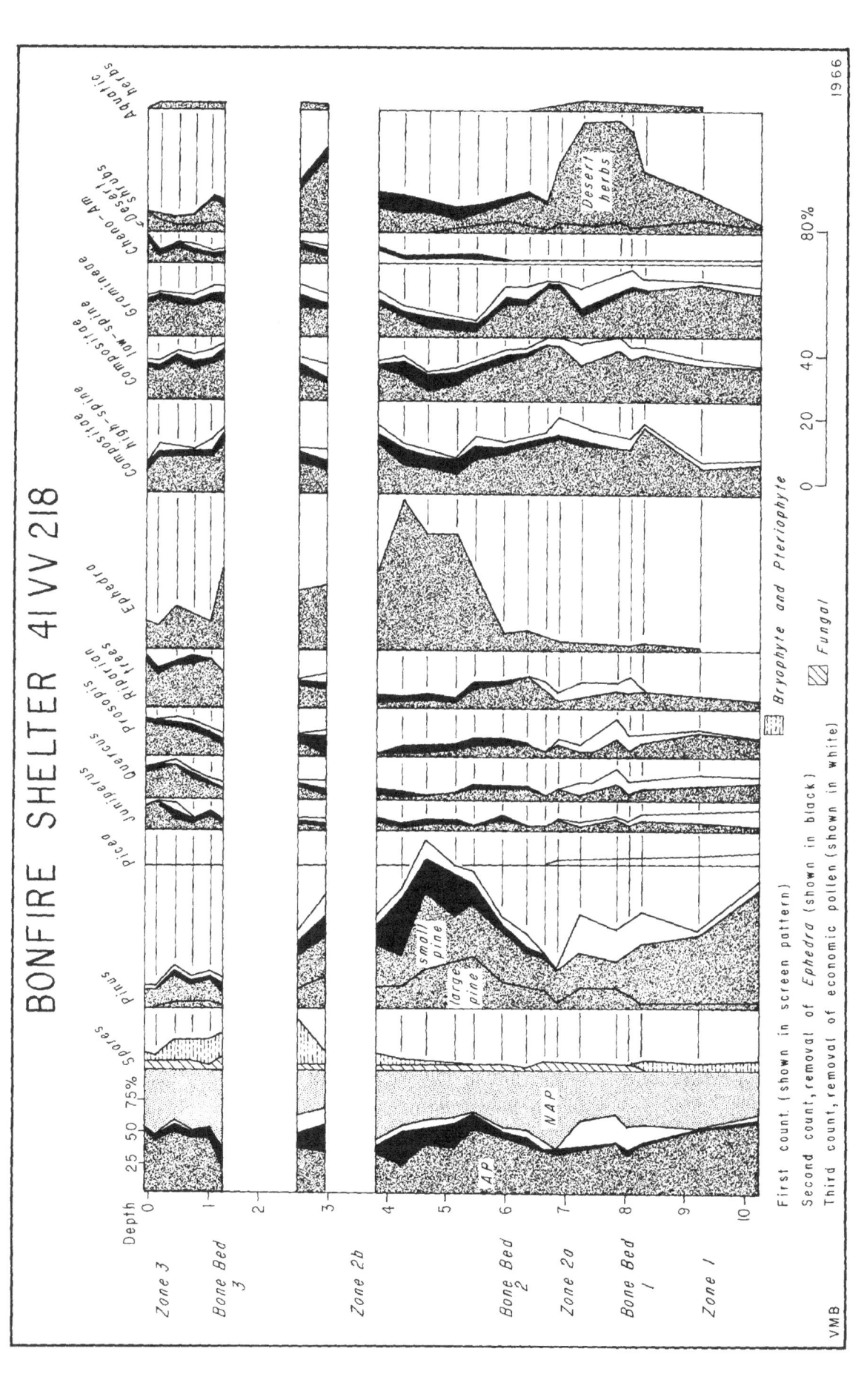

First count.(shown in screen pattern)
Second count,removal of *Ephedra* (shown in black)
Third count,removal of economic pollen (shown in white)

▦ Bryophyte and Pteriophyte
▨ Fungal

1966

VMB

172

sporadic non-arboreal pollen included <u>Larrea</u>, <u>Fouquieria</u>, <u>Artemisia</u>, Malvaceae (c.f. <u>Sphaeralcea</u>), Onagraceae (c.f. <u>Oenothera</u> and <u>Gaura</u>), Polemoniaceae, <u>Kallestromia</u>, <u>Erio-gonum</u>, <u>Agave</u>, and <u>Cucurbita</u>, but <u>Ephedra</u> is strangely absent.

Bone Bed 1, Zone 2a, and Bone Bed 2 are characterized by percentages of <u>Pinus</u>, <u>Juniperus</u>, <u>Quercus</u>, and <u>Prosopis</u> equal to or less than that of the modern pollen rain. Pollen of aquatic herbs and riparian trees, including <u>Alnus</u>, increases in relative abundance in this zone. Fossil Cheno-Am pollen remains much below modern levels, but <u>Ephedra</u> pollen appears and achieves a relative abundance equal to that of the modern pollen rain. Gramineae and Compositae pollen are generally equal to or greater than the modern pollen records except in Zone 2a where Compositae and Gramineae in at least two samples drop to below average percentages. This drop is spurious as it coincides with an abrupt rise of two probable economic types, Onagraceae and Malvaceae, to 25% and 15% respectively. The record of large entomorphilous non-arboreal types, which may tentatively be considered probable economic types, differs in Profiles B and C. In Profile B <u>Agave</u> pollen, which was common in Zone 1, is lacking in Bone Bed 1, Zone 2a, and Bone Bed 2; however, pollen of <u>Dasylirion</u>, <u>Nolina</u>, <u>Yucca</u>, and sedge pollen replace it, the latter type being most abundant in the bone beds. In Profile C (Figure 28), <u>Agave</u> pollen is very abundant in Bone Bed 2 as is the pollen of the Onagraceae. Cactaceae pollen is two to three times more abundant and Malvaceae pollen only about one-half as abundant in Profile C as in Profile B. Riparian tree pollen likewise differs in Profiles B and C, as in the former profile this category increases in Bone Bed 2 but in the latter profile it is much below the levels found in the modern pollen rain.

Zone 2b is characterized by extraordinarily high relative abundance of <u>Ephedra</u> pollen and the surprising appearance of the <u>Ephedra</u> <u>nevadensis</u> type as well as <u>Artemisia</u> of the <u>A. filifolia</u>. The lower levels of Zone 2b are characterized by high percentages of <u>Pinus</u> pollen, but in the upper levels this type declines in relative abundance. This change in pine frequency precedes the maximal percentages of <u>Ephedra</u> <u>nevadensis</u> and <u>Artemisia</u> <u>filifolia</u> types and also the occurrence of scattered hearths, charcoal of which has been dated about 7240 ± 220 B.P. (Sample Tx-152; Pearson, <u>et al.</u>, 1965). Below these hearths and in association with the pine and <u>Ephedra</u> peak, Gramineae pollen declines to a relative abundance well below the modern record; Cheno-Am pollen increases slightly; Compositae pollen, while fluctuating to below average values, remains generally comparable to the modern pollen record, and most large entomophilous non-arboreal pollen types decline in relative abundance or disappear except for Cactaceae pollen whose relative abundance increases in Profile B only. Above these hearths Juniperus, Prosopis, and riparian tree pollen increase slightly, but Pinus and Quercus pollen remain low and relatively unchanged, while Ephedra pollen declines only slightly, remaining well above modern levels. Compositae, Gramineae, and Cheno-Am pollen increase to percentages equal to or greater than the modern pollen rain and in general quite comparable to those recovered from Bone Bed 3. Spores of Selaginella, Bryophytes, and various ferns are quite common following the maxima of Ephedra and Pinus.

Bone Bed 3 and Zone 3 (including a conspicuous "fiber layer" of perishable artifacts) are characterized by variable pine percentages, which in Profile C are equal to those of the modern pollen rain, while in Profile B they are generally less than those of the modern records. Juniperus pollen increases in relative abundance in Zone 3, but the change observed exceeds the variation noted in the modern record only slightly. Quercus, Prosopis, Celtis, Gramineae, and Ephedra pollen are all in proportions exceeding those of the modern pollen rain; but Compositae and Cheno-Am pollen percentages are essentially similar to those of the modern record. Ephedra and large trees of Prosopis and Celtis are near the entrance of the shelter so the excessive quantities of these types are not unexpected, and it is entirely possible that due to the condition of preservation some misidentified grains of Prosopis have been included in the Quercus category. The relative abundance of the arboreal types is of significance not so much in comparison to the modern record but in comparison to preceding fossil levels where these types were relatively much more scarce, except, of course, for pollen of Pinus, which was much more abundant.

COMPARISONS

There are many Southwestern pollen records which could be compared with that from Bonfire Shelter, but the present comparison will be limited to four other sites from the Amistad Reservoir; Centipede and Damp Caves (Johnson, 1963), Eagle Cave (McAndrews and Larson, herein), and the Devil's Mouth Site (Bryant, herein). The longest and most complete fossil pollen records have been obtained from Bonefire

FIGURE 26.   Fossil pollen record at Profile B, Bonfire
             Shelter, arboreal pollen (N=100). Percentage
             of broken pine grains is shown to left of
             pine curve. Pine curve is divided into per-
             centage of large and small grains.

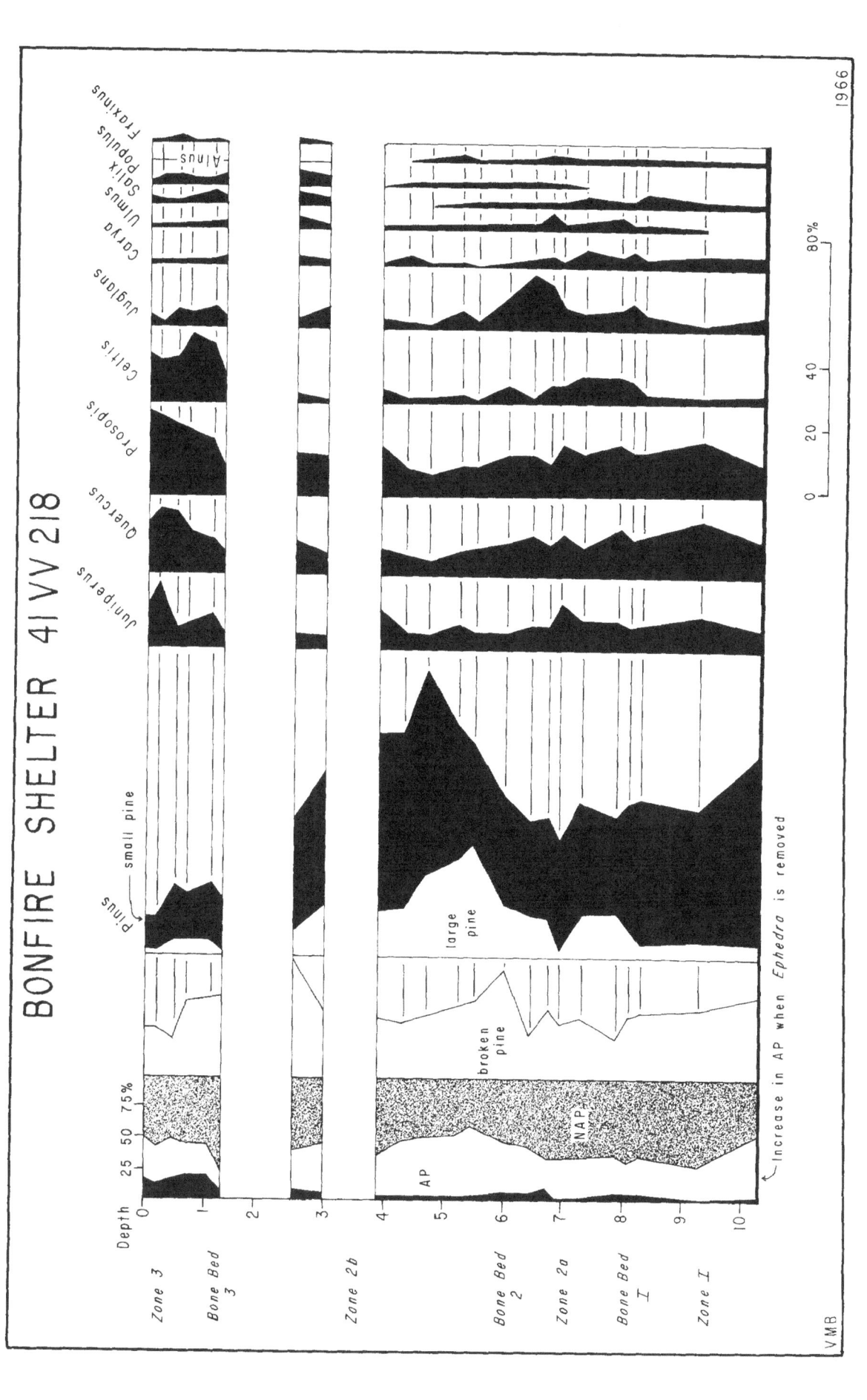

BONFIRE SHELTER 41 VV 218

1966

VMB

FIGURE 27.   Non-arboreal fossil pollen record at
             Profile B.

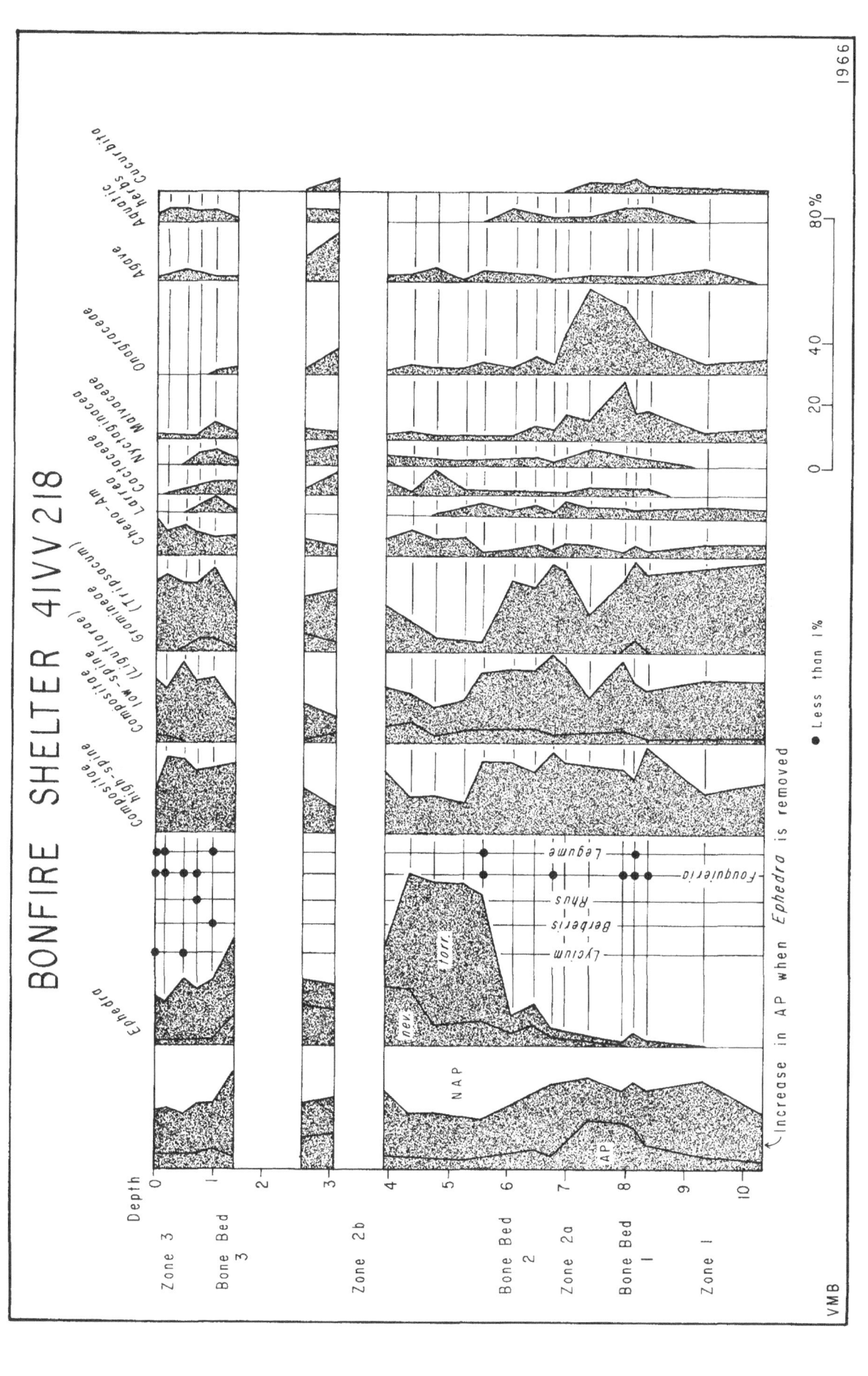

BONFIRE SHELTER 4IVV218

1966

FIGURE 28.   Fossil pollen record at Profile C, Bonfire
              Shelter.

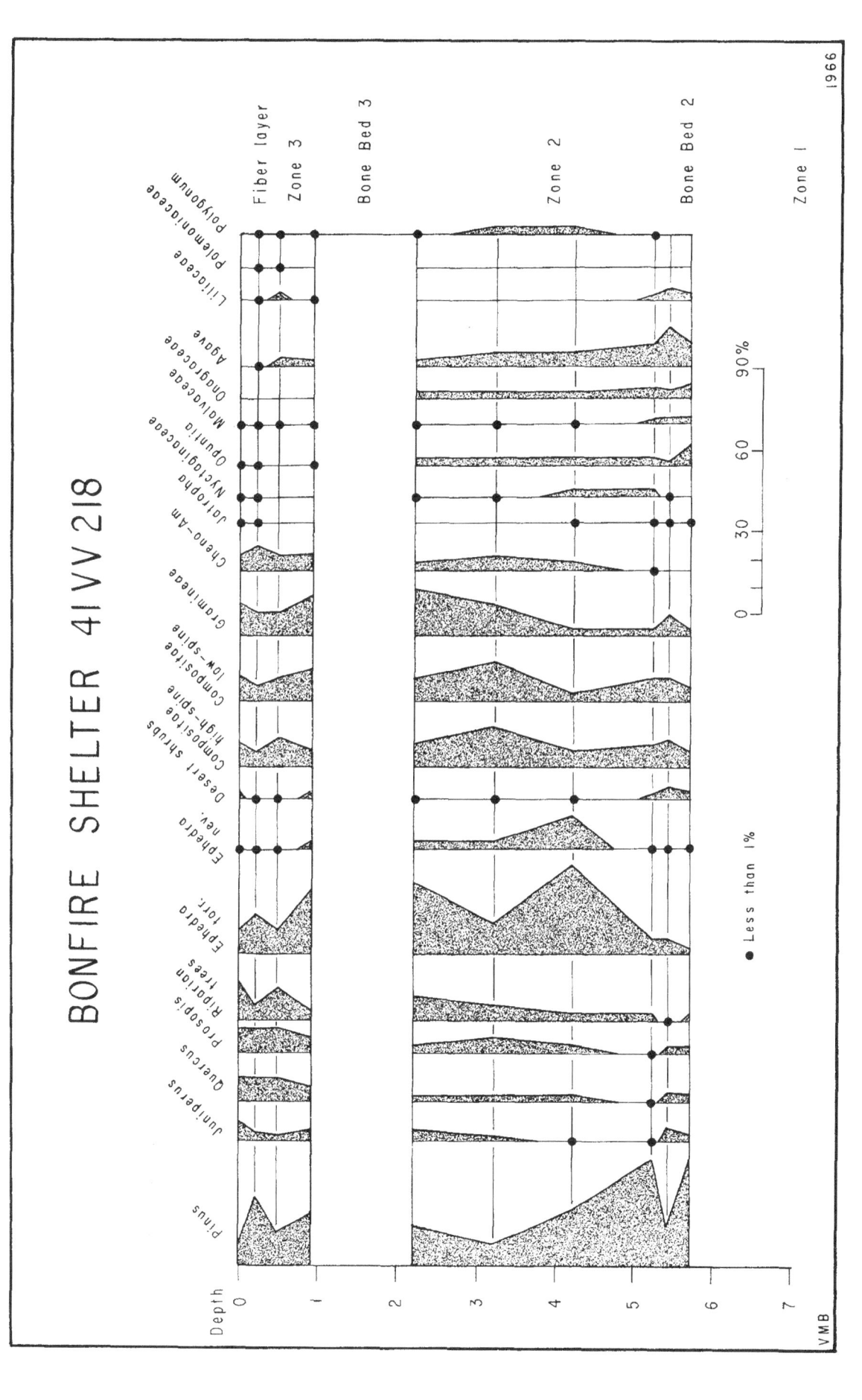

BONFIRE SHELTER 41VV218

1966

Shelter and Devil's Mouth sites where sediments containing
Paleo-Indian artifacts have been found, and at least in
the case of Bonfire Shelter there is also a record of extinct fauna. In both sites the deepest excavated levels
are characterized by percentages of _Pinus_ pollen exceeding
those of my modern record (Figure 24), and in the suc-
ceeding levels there is a particularly pronounced increase
of this type. The deepest levels of Bonfire Shelter also
exhibit a decrease of _Pinus_ pollen, but this depression
seems to be an artifact of restraint imposed by economic
pollen. Although culturally sterile and undated, the
lowest levels of Eagle Cave also seem to possess a record
that may parallel that of the Devil's Mouth and Bonfire
Shelter sites. Closely following or partially coincident
with the second pine peak is an abrupt increase in _Ephedra_
pollen. In the pollen record from Bonfire Shelter the
percentages of this type so greatly exceed modern records
collected in close proximity to _Ephedra_ plants that a
selective concentration by man or other agency must be
assumed; nevertheless, some evidence of increased relative abundance of _Ephedra_ pollen is demonstrated in Eagle
Cave, Centipede and Damp Caves, as well as in the Devil's
Mouth site. In Bonfire Shelter there is evidence that the
second pine peak was beginning by 10,230 B.P. (Tx-153;
Pearson, _et al._, 1965), and at Eagle Cave it was ending
by about 8650 B.P. The _Ephedra_ peak, or actually a minor
depression of this type, corresponds with the occurrence
of "intermediate horizon" hearths, charcoal of which has
been dated at 7240 B.P. (Tx-152; Pearson _et al._, 1965).

The _Ephedra_ peak fluctuates considerably but remains
well above modern levels until about 2310 or 2810 B.P.,
when _Pinus_, _Juniperus_, _Quercus_, _Prosopis_, _Celtis_, Compositae, and Gramineae pollen equal or exceed the values of
these types in the modern pollen rain. Between 7240 B.P.
and 2310 B.P. there is at least one brief interval during
which _Ephedra_ decreases to very nearly modern levels, and
a slight increase in _Juniperus_, _Quercus_, _Prosopis_, riparian
tree, Compositae, and Gramineae pollen occurs. Most of
these types decline briefly in relative abundance with the
second major peak of _Ephedra_ that occurs just before 2310
or 2810 B.P., but increase in relative abundance in the
uppermost sediments of the cave. Similar trends are not
apparent in the pollen record from the alluvial terraces
at Devil's Mouth Site nor particularly from Eagle Cave,
Centipede Cave, or Damp Cave.

OBSERVATIONS AND CONCLUSIONS

An extremely sensitive pollen record is preserved in the sediments of Bonfire Shelter, and analysis of these sediments has revealed a long history of vegetation change. Although plants characteristic of desert habitats have occupied the area about Bonfire Shelter during the past 10,000 years, there have been conspicuous changes in the composition of past plant communities that reflect climatic variations. Five difference periods may be recognized on the basis of total AP-NAP fluctuations and changing relative abundance of either AP or NAP types in pollen Profile B. Exclusion of Ephedra and/or "economic" pollen types reinforces the distinctiveness of some of the periods but tends to obscure others (Figures 25,26, & 27).

1. The lowest undated levels of Zone 1 are characterized by high AP values and a high relative abundance of grass pollen but very low relative abundance of pollen from riparian or desert plants, suggesting a warm steppe or savanna-woodland environment with scattered mesquite, oak, juniper, and pine. No modern pollen samples obtained as yet from the Amistad Reservoir or Edward Plateau exhibit the features of the lowest sample thus far obtained from Zone 1 (10.3 feet), but the next sample higher (9.3 feet) is quite similar to samples obtained from oak-juniper woodland with widely scattered pinyon pine except, of course, for the 1-2 grains consistently recovered from each of the lower zones of Bonfire Shelter and suggesting that the expansion of the mesic communities from the Edwards Plateau to Mile Canyon was coincident with the well documented expansion of spruce and pine communities further to the north and west (Hafsten, 1961; Hevly, 1964; Clisby and Sears, 1956; Martin, 1963).

2. Between 9.3 and 6.0 feet AP pollen diminishes in relative abundance, while the pollen of NAP increases, particularly that of certain desert shrubs and herbs, including Ephedra, which makes it initial appearance in the record at a depth of 8.3 feet, just inches below Bone Bed 1. Many pollen types exhibit decreased relative abundance at depths of between 6.9 and 8.3 feet due to the exceptionally great quantities of "economic pollen." It is interesting to note that evidence of human utilization of the cave during this lower or earlier interval is scarce, consisting thus far of broken bones of disarticulated skeletons of camel, horse, bison, and mammoth, limestone spalls thought to have been used in smashing the bones to recover marrow, and a few flecks of charcoal in Bone Bed 1 at a depth of 7.9 feet (Dibble, 1965).

175

Exclusion of the "economic" plant category demonstrates that in reality the high AP values and environment of the preceding two levels actually persist up to a depth of 7.3 feet, and the period of decreased AP extends only from a depth of 7.3 to 6.0 feet.

The environment of the concealed high AP period associated with Bone Bed 1 may be considered to be identical or only slightly more xeric than the preceding levels of Zone 1. The low AP interval of Zone 2a appears palynologically similar to the modern pollen rain of the Bonfire Shelter area today except in the high grass percentages. This interval probably represents a desert grassland rather than a desert and is associated with Bone Bed 2 which has been dated at 10,230 ± 160 B.P. There is thus a striking difference in the vegetation of the Bonfire Shelter area at the time that camel, horse, bison, and mammoth (Bone Bed 1) may have been hunted from when the extinct bison (Bone Bed 2) was slaughtered. It is interesting to note that both intervals are characterized by increased relative abundance of grass and herb pollen and also by slightly increased pollen of riparian trees.

3. Between 6.0 and 4.3 feet AP values reach maximal proportions and so do Ephedra percentages. Exclusion of fossil Ephedra pollen, which exceeds values for this type in the modern pollen of soil surfaces collected in close proximity to Ephedra, results principally in increase of pine, Compositae, and "economic" pollen. The combination of high pine and Ephedra percentages is unnatural unless regarded as a cave phenomena in which these anemophilous pollen types have been selectively concentrated, while others such as low-spine Compositae, Gramineae, and Cheno-Ams have been excluded. This seems most unlikely, particularly in light of the high percentages of large entomophilous pollen also recovered, so Ephedra has been regarded as being concentrated by other means. In no other site is so much Ephedra pollen recovered, although it must be observed that other sites do exhibit an increase in this type, but with the Ephedra peak in no way coinciding with that of Pinus and definitely following it. This sequence is also suggested in Bonfire Shelter, as the pine maximum (particularly that of large pine) occurs before the Ephedra maximum, especially in Profile C. The interpretation of the phenomena of this and the following interval must await further studies currently underway.

4. From 4.3 feet to 1 foot below the surface of the sediments of this shelter is a stratum characterized by low arboreal pollen and very poor pollen preservation.

Ephedra, Gramineae, and economic pollen are the dominant NAP, but removal of *Ephedra* and economic pollen discloses a high AP record similar to that of Zones 1 and 2, but quite different from that of the underlying Bone Bed 2 or Zone 2b and from that of the overlying Bone Bed 3 and Zone 3. Attempts are currently being made to complete the pollen record of this stratum, and interpretation is best delayed until more data is available. In most respects, the samples of this interval are like the modern pollen rain of the Bonfire Shelter area.

5. The upper foot of sediment contains a pollen record unlike the modern records in its very low relative abundance of pine pollen and its high relative abundance of *Celtis*, *Prosopis*, *Quercus*, and *Juniperus*. The high relative abundance of *Celtis* and *Prosopis* is not unexpected as larger trees of both grow at the entrance of the shelter, but as yet I have no explanation for the excessive quantities of *Juniperus* and *Quercus*, reflecting an apparent increase in these plants sometime in the last 250 to 550 years. It is interesting to note that evidence of more mesic conditions in the last few centuries is not unique to this cave but was also detected in sediments from Lake Texcoco in the Valley of Mexico (estimated at post 900 A.D.) by Sears (1952) and in many pollen studies of sediments from pueblo sites in Arizona and New Mexico where more mesic intervals have been dated at 300-700 to 1000-1200, 1300-1500, and 1700-1900 A.D. (Schoenwetter and Eddy, 1964; Hevly, 1964).

"Economic Pollen"

The principal cultural horizons of Bonfire Shelter are each characterized by specific assemblages of economic types. Bone Bed 1 has a high relative abundance of Malvaceae (c.f. *Sphaeralcea*), Onagraceae (c.f. *Oenothera* and *Guara*), Nyctaginaceae, Cyperaceae, *Cucurbita*, and *Tripsacum*. Bone Bed 2 has the same types except *Cucurbita* and *Tripsacum*; however, they are much less abundant, although well above modern levels. The "intermediate horizon" has abundant Nyctaginaceae, *Opuntia*, *Typha*, *Cucurbita*, *Agave*, and *Tripsacum*, but the relative abundance of Malvaceae and Onagraceae is similar to that found in Bone Bed 2. Bone Bed 3 and the fiber layer of Zone 3 are characterized by the scarcity or absence of Onagraceae, Malvaceae, Nyctaginaceae, and *Cucurbita* and the presence of *Tripsacum*, *Opuntia*, *Agave*, *Typha*, and Cyperaceae. *Tripsacum* and *Cucurbita* are present only in all cultural levels but Bone Bed 2; hence, Bone Bed 2 represents a different sort of utilization of the shelter from that of the other cultural horizons.

Suggestions for Future Work

1. Pollen Profile B presently lacks only a few samples of being a continuous record of the last 10,000 years, and an effort should be made to fill in these gaps.

2. Pollen Profile B should be extended to bedrock of the shelter, necessitating an unknown amount of further excavation in this particular pit, but the extension of the unique pollen record of this shelter would provide an opportunity of elucidating the effect of late-Pleistocene climates in this arid area and the role they may have played in the origin and evolution of the biota of southeastern North America (Braun, 1955; Deevey, 1949; Graham, 1965; Hutchinson, Patrick, and Deevey, 1958; Martin and Harrell, 1957).

3. Additional radiocarbon dates are badly needed but particularly for Bone Bed 1 and its associated pollen record.

4. Additional pollen profiles should be obtained from the shelter and the techniques employed at Pollen Profile B extended to them. Repeated sampling of each stratigraphic unit would assure recovery of the most complete record possible and verify correlations from one part of the shelter to another.

5. Palynalsystematic studies of critical entomophilous types from taxonomically or ecologically restricted groups would permit a more detailed reconstruction of past environments if the geographic distribution of the plants producing these pollen types was ascertained more critically.

6. Correlations of the findings of this shelter with other archeological sites in the Amistad Reservoir and pollen records from elsewhere in Texas (Potzger and Thorp, 1947, 1954) and the Southwest.

POLLEN ANALYSIS OF EAGLE CAVE

John H. McAndrews and Donald A. Larson

Eagle Cave is one of the archeological sites in the Amistad Reservoir excavated by the Texas Archeological Salvage Project. This site was one of the four sites chosen for palynological study.

A total of twenty-three soil samples representing all strata in the excavation were collected by Dr. John H. McAndrews and Vaughn M. Bryant, Jr. Each sample was collected from the cleaned vertical walls of the excavation with a trowel and placed in labeled polyethylene bags. Extreme care was taken to avoid contamination.

The samples were processed in the Palynological Laboratory of the Department of Botany, The University of Texas, using a digestion process very similar to the one described in the Devil's Mouth Report (Bryant, report herein). Pollen in significant amounts for analysis was obtained from eight of the twenty-three samples (Table 8). However, the eight samples yielding sufficient pollen represent an adequate record for Period II and the underlying culturally-sterile deposits. Samples representing time Periods III and IV were deficient of pollen probably because of oxidation by fire. Samples representing Periods V through VII were discarded because of evidence of significant strata disturbance by man.

Following digestion of soil and non-polleniferous plant fragments pollen residues were incorporated in silicone fluids of 2,000 cs viscosity and slides were made. Pollen counts were made by McAndrews at 400 and 1000 diameters. Identifications of pollen types were based upon material in the Amistad Pollen Reference Collection maintained by the Palynological Laboratory at The University of Texas.

The results of this study for the purposes of this report are shown in a list of identified types (Table 9) and a pollen diagram (Figure 29). Working with the time sequences outlined by Dr. Dee Ann Story it was found, as discussed in Appendix A of the Devil's Mouth report, that the pollen diagram is in basic agreement with the diagrams from Bonfire Shelter and the Devil's Mouth Site. It can be pointed out that three pollen types encountered in the samples from time Period II represent possible demonstration of plant use by primitive man. These types are Typha, Dasylirion, and Agave. Interestingly, these three types were encountered in Bonfire Shelter during the equivalent time period but are absent from the pollen record of the Devil's Mouth Site (Hevly, herein; Bryant, herein).

179

FIGURE 29.   Pollen diagram, Eagle Cave.

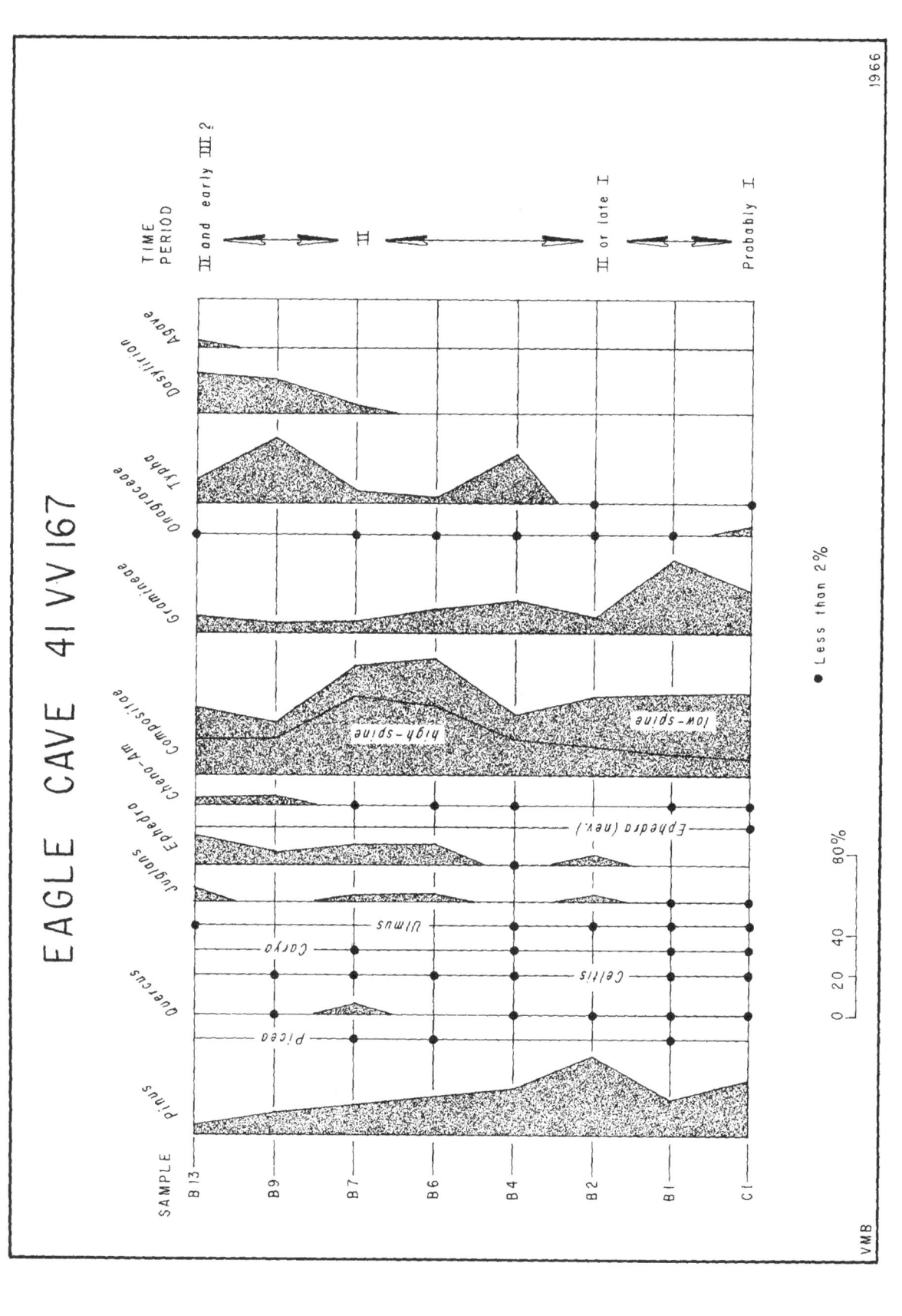

EAGLE CAVE 41 VV 167

1966

VMB

TABLE 8.   SOIL SAMPLES FROM EAGLE CAVE

Column B

Sample	Depth Below Surface	Stratum
1	8.8	Sterile stratum
2	7.8	" "
3	7.7	" "
4	7.3	V
5	7.3	V
6	7.4	V
7	6.9	V
8	6.5	V
9	6.1	IV
10	5.8	IV
11	5.4	III
12	5.1	III
13	5.6	III
14	3.5	IId
15	3.5	IId
16	2.8	IIc
17	2.2	IIb
18	1.0	IIa
19	1.5	IIa
20	.5	I
21		surface

Column C

Sample	Depth Below Surface	Stratum
1	9.5	Sterile stratum
2	10.0	Sterile stratum

TABLE 9. EAGLE CAVE POLLEN COUNT

Stratum	Sterile Layer below V			V			IV	III
Sample No. Pollen Types	C1	B1	B2	B4	B6	B7	B9	B13
Pinus (undif.)	63	35	110	48	41	33	23	14
Pinus (diplox. t.)	-	-	1	3	6	7	3	-
Pinus total	63	35	111	51	47	40	26	14
cf. Picea	-	*	-	-	1	1	-	-
Quercus	1	1	3	1	-	11	4	-
Celtis	1	3	-	2	1	1	1	1
Carya	2	1	-	3	-	1	-	-
Ulmus	1	5	4	-	-	-	-	2
Juglans	3	4	8	-	8	8	-	15
Ephedra (nev. t.)	1	-	-	-	-	-	-	-
Ephedra (tor. t.)	-	-	12	7	20	20	9	18
Rhamnaceae	-	-	1	-	-	-	-	-
Fraxinus	-	-	-	-	2	1	1	1
cf. Maclura	-	-	-	-	1	-	-	1
cf. Lonicera albiflora	-	-	-	-	2	-	-	-
cf. Acacia	-	-	-	-	-	1	-	-
cf. Diospyros	-	-	-	-	-	1	-	-
cf. Forestiera	-	-	-	-	-	1	-	-
Koeberlinia t.	-	-	-	-	-	1	-	-
cf. Acer	-	-	-	-	-	4	-	-

TABLE 9 (cont'd)

Stratum	Sterile Layer below V			V			IV	III
Sample No. / Pollen Types	C1	B1	B2	B4	B6	B7	B9	B13
Morus	-	-	-	-	-	-	2	-
Alnus	1	*	4	1	-	-	1	-
Cheno-Ams	1	1	-	1	2	4	8	10
Liguliflorae	-	*	-	2	-	-	-	-
Compositae (hi-spine)	19	25	32	33	86	99	38	44
Iva ambrosiaefolia t.	26	12	41	10	38	26	8	9
Ambrosia	39	59	11	14	17	4	5	7
Iva texensis t.	-	4	4	-	-	3	-	-
Xanthium	-	-	1	-	-	1	-	-
Artemisia	6	8	28	-	4	2	-	2
Compositae total	90	88	117	59	145	135	51	62
Gramineae	40	96	21	34	11	15	15	30
Onagraceae (undif.)	4	-	5	2	5	3	-	1
cf. Gaura	1	1	1	-	-	-	-	-
cf. Jussiasa	1	-	-	-	-	-	-	-
cf. Oenothera lamp.	-	1	-	-	-	-	-	-
cf. Cactaceae	2	-	-	-	-	-	-	-
Sphaeralcea t.	2	1	-	1	-	-	3	1
Typha domingensis t.	3	-	2	48	5	9	60	32
Cruciferae	-	-	-	-	1	-	-	-
Erodium	-	2	-	-	-	-	-	-

TABLE 9. (cont'd)

Stratum	Sterile layer below V			V			IV	III
Sample No. / Pollen Types	C1	B1	B2	B4	B6	B7	B9	B13
Dasylirion	-	-	-	-	-	7	35	40
Polypodiaceae (trilete)	-	3	1	4	4	-	3	2
cf. Rosaceae	-	-	1	-	-	-	-	-
cf. Phacelia	-	-	-	1	-	-	-	-
Dalea t.	-	-	-	-	-	-	2	13
Gilia	-	-	-	-	-	-	1	-
cf. Agave lech.	-	-	-	-	-	-	-	12
cf. Oxybaphus	-	-	-	-	-	-	-	1
cf. Yucca	-	-	-	-	-	-	-	1
TOTALS	218	262	291	214	256	263	221	247

t. = type

* = trace

POLLEN ANALYSIS OF THE DEVILS ROCKSHELTER

Vaughn M. Bryant, Jr.

This report consists of the preliminary study and results of a pollen analysis conducted on soil samples collected at the Devils Rockshelter Site (41 VV 264) near the lower end of the proposed Amistad Reservoir. The site is a narrow, open rockshelter at the foot of the limestone bluffs located on the northeastern side of the entrenched Rio Grande valley one-quarter of a mile downstream from the mouth of the Devils River (Figures 3, 20, a).

The site was discovered in 1964 by members of the Texas Archeological Salvage Project during a survey of sites in the Amistad Reservoir (Prewitt, 1966). The survey crew reported that the surface of the site had been badly mixed and disturbed but recommended that the site be tested. In March and April of 1965, a field crew from the T.A.S.P., under the leadership of Elton Prewitt, returned to the site for two weeks of limited investigation. Their excavations reached a maximum depth of fifteen feet and uncovered nine distinct strata (Figure 20,b). The deposits contained a variety of lithic artifacts (i.e., a pecked and grooved stone, projectile points, bifacial blades, core tools, a stone drill, gravers, scrapers, burins) and numerous burned rock fragments (Prewitt, 1966).

The soil samples analyzed in this report were collected on May 8, 1965, by Dr. Donald A. Larson and the writer. A total of fifteen samples were taken at six-inch intervals from the cleaned vertical walls of the largest excavation pit (Table 10). Each sample was collected with a clean trowel and placed in an uncontaminated polyethylene bag which was quickly sealed. Each bag was then labeled with both the archeological site number and the pollen sample's provenience before being placed in separate paper bag for storage.

Samples (numbers 1, 4, 7, 10, 13) representing five of nine strata were analyzed in this preliminary study. Each sample was processed in the Palynological Laboratory at The University of Texas, using digestion techniques developed during the pollen analysis of the Devil's Mouth

185

Site (see report herein). Essentially, four basic procedures are involved: 1) washing in concentrated hydrochloric acid, 2) bathing into a 10% solution of potassium hydroxide, 3) treating with concentrated hydrofluoric acid, and 4) acetylation. Following digestion, the pollen residues were mixed with 2000 cs silicone oil and placed in separately labeled 5 ml. shell vials.

After examining a number of microscopic slides prepared from each of the five processed samples, the writer found a total of only seven recognizable pollen grains, seventeen pollen and spore fragments, and two unidentifiable crushed pollen grains (Table 11). Less than one-quarter of a mile away, the Devil's Mouth Site yielded a continuous pollen record extending from the upper part of Period I to the present. No further digestions and pollen analytical studies were conducted on material from Devils Rockshelter.

Further studies of soil chemistry at the Devils Rockshelter Site may yield clues as to the reasons why these samples contain such low frequencies of pollen. However, in the opinion of this writer, any future attempt to develop a full pollen record for the site would be an inappropriate use of time and equipment.

TABLE 10.

SOIL SAMPLES FROM THE DEVILS ROCKSHELTER SITE

Sample	Zone	Depth (in feet below surface)
15	Surface	Surface
14	Zone VII	.75
13	Zone VI	1.25
12	Zone VI-V	1.75
11	Zone V	2.25
10	Zone V-IV	2.75
9	Zone IV	3.25
8	Zone III	3.75
7	Zone III	4.25
6	Zone II	4.75
5	Zone II	5.25
4	Zone II	5.75
3	Zone II-Ie	6.25
2	Zone Ie	6.75
1	Zone Ie	7.25

TABLE 11

DEVILS ROCKSHELTER POLLEN COUNT

Sample No.	Pollen Type	Number of Pollen Grains
13	Gramineae	1
	Unknown	1
	Fragment	3
10	Cheno-Am	1
	Opuntia sp.	1
	Unknown	1
	Fragment	6
7	Compositae (high-spine)	1
	Pinus sp.	1
	Fragment	7
4	None	
1	Opuntia sp.	2
	Fragment	1

188

BOTANICAL STUDIES IN THE AMISTAD RESERVOIR, A SURVEY

Donald A. Larson and Vaughn M. Bryant, Jr.

Modern Flora

The preliminary collection of contemporary plant taxa in the Amistad region has provided the framework for an understanding of the flora in this xeric region. However, the results outlined in this report only represent a beginning; a more intensive study is needed.

Additional studies are needed in order to determine the full complement of plant species growing on both sides of the Rio Grande. Furthermore, a greater understanding of plant ecological zonations (i.e., plateau, protected canyons, river terraces) within the Amistad Reservoir area cannot be understood until additional modern collections can be made and identified. Further collection and characterization of the contemporary flora of this region will also provide essential data needed in developing a more comprehensive pollen reference collection of the region.

Plant Macrofossil Studies

The results outlined in this report prove that analyses of plant macrofossil remains from archeological sites are valuable tools for reconstructing cultural patterns of primitive man and can provide essential correlative date for plant microfossil studies from the same sites. Only a portion of the existing plant macrofossils from archeological sites in the Amistad Reservoir area were analyzed in this report. These studies need to be expanded to include: 1) the remaining plant macrofossils not studied in this report, 2) plant materials currently being discovered in new archeological sites under excavation, and 3) plant material which will undoubtedly be recovered from new archeological sites planned for excavation in the near future.

Joint efforts by botanists and archeologists in the collection of plant macrofossils during actual periods of archeological excavations is considered essential. Furthermore, inter-disciplinary teamwork between botanists

189

collecting and identifying plant macrofossils and the
archeologists excavating the sites will provide each
discipline with firsthand information essential to the understanding of the significance of the available data.

Palynology

Palynological work deserves considerable expansion.
Pollen studies in the Amistad region have already provided
general guide lines in understanding the varying climatic
and vegetational conditions of the late Quaternary period.
However, future fossil pollen studies of archeological sites
currently under investigation in the Amistad region will
hopefully provide additional data as to the: 1) nature of
climatic and vegetational conditions prior to and during
periods when extinct bison 'herds were roaming the area,
and prior to that the Pleistocene mammals whose bones are
associated with the lowest bone bed of Bonfire Shelter,
2) climatic and vegetational conditions from the end of the
full-glacial period through to contemporary time, 3)
climatic conditions that triggered the severe erosional
interval (termination of Altithermal) during time Period V
and the nature of the vegetational changes during this
period, 4) economic plants used by primitive man and insights into his diet through examinations of human
coprolites found in archeological sites, 5) methods of
correlating the pollen records of terrace and rockshelter
sites, and 6) a means of relative dating for archeological
sites in the Amistad region through comparisons of their
fossil pollen records.

Considerable expansion of the Amistad Pollen Reference Collection and the pollen key is needed. Furthermore, intensive studies of modern pollen rains with regard
to the sources of anemophilous pollen will enhance the
interpretation of airborne pollen of woody plants in fossil
pollen rains.

Continued botanical studies in the Amistad Reservoir area will provide important data which can be coupled
with archeological and geologic information. Together they
can be used as a framework on which to build an understanding
of the paleoecology of southwest Texas during the late
Quaternary period. However, this important and essential
study must be continued in the immediate future since the
Amistad Dam is to be completed by 1968 and the resultant
lake will destroy all of the existing data.

INTRODUCTION TO ZOOLOGICAL STUDIES

Gerald G. Raun

Zoological field work was concentrated in three primary areas: investigation of the modern vertebrate fauna (restricted primarily to herpetofauna), analysis of vertebrate remains from archeological sites, and analysis of molluscan remains from archeological sites. This was a preliminary study, of short duration, and consequently has raised more problems than it has answered. Much remains to be learned and much field work remains to be done, but preliminary evidence indicates that the Val Verde County is a zoogeographically important area and deserves a concentrated study by workers in various disciplines.

Separate reports on the modern herpetofauna, the paleovertebrate fauna, and the paleomolluscan fauna are included.

PRELIMINARY REPORT ON THE HERPETOFAUNA OF

VAL VERDE COUNTY

Gerald G. Raun

Field work in Val Verde County, carried on during the summer of 1965 and periodically through the fall and winter, was primarily concentrated upon the herpetofauna. No particular effort was made to collect mammals, but some observations were made. Fish and birds were not collected. The collections made during this study were not extensive and the herpetofauna is far from perfectly known; however, certain conclusions can be drawn based upon the records reported here and other distributional data, acquired in the process of making an extensive survey of the distribution of Texas reptiles and amphibians. I have covered the literature quite thoroughly, and the range maps which I have plotted should be as accurate as possible with the existing information.

An analysis of the general zoogeographically affinities of the Val Verde herpetofauna can be made, but it is clear that extensive field work remains to be done to obtain specific ecological information. Basic vegetational data are now available (see report by Flyr, herein) and considerable effort should be devoted to detailed habitat and distributional studies. This should prove rewarding since Val Verde County is an area of faunal interchange and the small amount of work done so far has yielded some interesting information.

Val Verde County is situated at the confluence of three biotic provinces: Tamaulipan, Balconian, and Chihuahuan (Blair, 1950). The fauna is, accordingly, comprised of elements of all three.

SPECIES LIST

Scaphiopus couchi. Couch's spadefoot toad. Twelve specimens were collected from the following localities: 3.5 mi. N. Comstock, 6.9 mi. N. Comstock, 8.1 mi. N. Comstock, 8.3 mi. N. Comstock, 16.8 mi. N. Comstock, 19.7 mi. N. Comstock, 19.9 mi. N. Comstock, 17.2 mi. S.E. Comstock, 8.3 mi. N. Del Rio, 8.9 mi. N. Del Rio, 10.6 mi. N. Del Rio, 18.5 mi. N. Del Rio. Val Verde County is well within the range of S. couchi, but apparently the species has not been reported from the county before. Couch's spadefoot is a western species which ranges through most of western Texas east to Tarrant, Bastrop, and Refugio counties.

Scaphiopus hammondi hammondi. Western spadefoot toad. Two specimens were collected from 23.2 mi. N. Comstock and 18.1 mi. N. Del Rio. These specimens represent eastern marginal records for the species. It has been reported from Terrell County (Milstead, et al., 1950) to the west and Crockett County (Brown, 1950) to the north, but not previously from Val Verde County. The western spadefoot ranges through Trans-Pecos and the Texas Panhandle east to Armstrong, Crockett, and Val Verde counties.

Eleutherodactylus augusti latrans. Barking frog. One specimen was collected 15.4 mi. N. of Comstock. This represents the first record for Val Verde County, although the species has been reported from Terrell County (Milstead, et al., 1950; Scudday, 1965) some 70 miles farther west. The barking frog is a Balconian species, occurring primarily on the Edwards Plateau of Texas and reaching its western known limits in Terrell County.

Bufo speciosus. Texas toad. Eight specimens were collected from the following localities: 14.3 mi. S.E. Comstock, 8.4 mi. N. Del Rio, 8.8 mi. N. Del Rio, 18.8 mi. N. Del Rio, 20.3 mi. N. Del Rio, 12 mi. N.W. Del Rio, and 12.5 mi. N.W. Del Rio. Val Verde County is well within the range of this western species which occurs as far east as Dallas, Leon, and Aransas counties.

Bufo punctatus. Red-spotted toad. Nine specimens were collected from the following localities: 9.2 mi. W. Comstock, 11 mi. N.W. Del Rio (3), 12.5 mi. N. W. Del Rio, 13 mi. N.W. Del Rio, 14 mi. N.W. Del Rio, and 1 mi. E. Langtry (2). This western species ranges throughout western Texas, east to Dallas, Bexar, and Cameron counties.

Bufo valliceps. Gulf Coast toad. Three specimens were collected 18 mi. N.E. of Comstock. This Mexican species ranges widely through eastern Texas, reaching its western known limits in Terrell County (Milstead, et al., 1950).

Acris crepitans blanchardi. Blanchard's cricket frog. Five specimens were collected from the following localities: 18 mi. N.E. Comstock, 13 mi. N.W. Del Rio (2), 14 mi. N.W. Del Rio (2). Acris crepitans is an eastern species which ranges across Texas to Jeff Davis and Brewster counties.

Gastrophryne olivacea. Great Plains narrow-mouthed toad. Three specimens were collected 14 mi. N. W. Del Rio. This is a central plains species which ranges from eastern Texas west to Jeff Davis and Presidio counties.

Rana pipiens. Leopard frog. Eighteen specimens were collected from the following localities: 10.4 mi. S.E. Comstock, 18 mi. N.E. Comstock (3), 6.2 mi. N. Del Rio, 6.5 mi. N. Del Rio, 14 mi. N.W. Del Rio (4), 1 mi. W. Langtry, 1 mi. E. Langtry (7). The leopard frog, as the species is currently understood, ranges across the United States and is one of the most ubiquitous of Texas' frogs.

Rana catesbeiana. Bullfrog. No specimens were collected but one was photographed at a stock tank near the mouth of the Devils River. The bullfrog is an eastern species which ranges west across most of Texas; however, it has been widely introduced and its natural range is difficult to substantiate at this time. This constitutes the first record from Val Verde County.

Kinosternon flavescens flavescens. Yellow mud turtle. Three specimens were collected 0.2 mi. W. (2) and 0.4 mi. W. of the intersection of U.S. Highways 90 and 163. This constitutes the first record of this species from the county, although it is well within the known range. This western species ranges throughout western Texas east to Dallas, Milam, and Victoria counties.

Pseudemys concinna texana. Texas slider. One specimen was collected at the mouth of the Devils River. The Texas slider has not been previously reported from Val Verde County. This is an eastern species which ranges west in Texas to Culberson and Brewster counties.

Pseudemys scripta elegans. Red-eared turtle. Two specimens were collected from 13 mi. N.W. and 14 mi. N.W. of Del Rio. This is the first report of the red-eared turtle from Val Verde County, although this eastern species is found throughout most of Texas.

Gopherus berlandieri. Texas tortoise. One specimen was collected 20.4 mi. N. of Del Rio. Tanzer, *et al.* (1966) reported a specimen of *G. berlandieri* from 45 mi. N. of Del Rio and commented that this extension of range was suspect because of the tendency for this species to be taken as pets, transported about the country, and released. I would be inclined to disregard the specimen reported here were it not for the existence of the second specimen. These records extend the range of the species northwest from Maverick County (Brown, 1950) the previous westernmost locality. G. berlandieri enters southern Texas and ranges north to Val Verde, Bexar, and Aransas counties. Its known range almost exactly defines the Tamaulipan Biotic Province as mapped by Blair (1950).

Trionyx spinifer emoryi. Texas spiny softshell. One specimen was collected 1 mi. W. of Langtry, but numerous softshells were seen in the Devils, Pecos, and Rio Grande rivers. The spiny softshell is an eastern species which ranges across Texas, excluding the western Panhandle. This may be the most abundant aquatic turtle in the Amistad area.

Coleonyx variegatus brevis. Banded gecko. Seven specimens were collected from the following localities: 12 mi. W. Comstock, 11 mi. N.W. Del Rio, 11.3 mi. N.W. Del Rio, and 1 mi. E. Langtry (4). The banded gecko, which is common in Val Verde County, ranges across Trans-Pecos, along the southern edge of the Edwards Plateau to Bexar County, and southeastward along the Rio Grande to Hidalgo and Kleberg counties. It is a western species.

Crotaphytus collaris. Collared lizard. Five specimens were collected from the following localities: 2 mi. S.E. Comstock, 13 mi. N.W. Del Rio, 14 mi. N. W. Del Rio, and 3 and 5 mi. W. of the Pecos River on U.S. Highway 90. The collared lizard is a western species which ranges east in Texas to the eastern edge of the Edwards Plateau, Limestone, and Dallas counties. It is a fairly abundant lizard in the Amistad area.

Holbrookia texana texana. Greater earless lizard. Eighteen specimens were collected from the following locali-ties: 1 mi. W. Comstock, 12 mi. W. Comstock, 12 mi. N.W. Del Rio, 13 mi. N.W. Del Rio (2), 14 mi. N.W. Del Rio, 1 mi. E. Langtry, 8.5 mi. E. Langtry, mouth of the Devils River (2), 1.5 mi. W. Devils River and 1 mi. N. Rio Grande (2), 3 mi. W. Devils River (3), Rio Grande 3 mi. upstream from mouth of the Pecos River, 0.5 mi. S.W. Pecos River Bridge (2). H. texana is a western species which ranges east in Texas to Dallas County, the eastern edge of the Edwards Plateau, and Hidalgo County. It is quite common in the study area.

Sceloporus merriami merriami. Merriam's canyon lizard. Twenty-seven specimens were collected from the following localities: 12 mi. W. Comstock (2), 18 mi. N.E. Comstock, 30 mi. N. Comstock (2), 32 mi. N. Comstock, 11 mi. N.W. Del Rio (2), 12 mi. N.W. Del Rio (3), 12.5 mi. N.W. Del Rio (3), 13 mi. N.W. Del Rio, 14 mi. N.W. Del Rio (2), 1 mi. E. Langtry (6), near the mouth of the Devils River, new Devils River Bridge, Rio Grande 3 mi. upstream from the mouth of the Pecos, Pecos River Bridge. This species has a restricted range in western Texas and adjacent Mexico. It ranges in Texas from Presidio County east to Crockett and Val Verde counties. This is the nearest to an endemic species of lizard found in the area and it is very abundant, particularly in the canyons.

Sceloporus olivaceus. Texas spiny lizard. Two specimens were collected from 14 mi. N.W. Del Rio and 2 mi. S.E. of the old Devils River bridge. S. olivaceus is found in central Texas, east to Smith, Nacogdoches, and Harris counties and west to Brewster County. An isolated population is present in El Paso County and adjacent New Mexico. This lizard is apparently uncommon in Val Verde County and was found only along wooded stream courses.

Sceloporus poinsetti. Crevice spiny lizard. Eight specimens were collected from the following localities: 13 mi. N.W. Del Rio (5), 0.2 mi. S.W. intersection Highway 90 and 163 (2), and Devils River 1 mi. upstream from the mouth. A western species ranging through Trans-Pecos Texas and across the Edwards Plateau to Brown and Bexar counties, the crevice spiny lizard is common on the canyon walls but is difficult to collect.

Sceloporus undulatus consobrinus. Southern prairie lizard. Six specimens were collected from the following localities: 1.5 mi. S. Comstock, 1.7 mi. S.E. Comstock, 12 mi. W. Comstock, 24 mi. N.E. of Comstock, 7.8 mi. E. Langtry, and the mouth of the Devils River. This is a wide ranging species which is found throughout Texas. It appears uncommon in Val Verde County and is apparently restricted to wooded areas primarily.

Urosaurus ornatus ornatus. Eastern tree lizard. Eleven specimens were collected from the following localities: 12 mi. W. Comstock (4), 1 mi. E. Langtry, mouth of the Devils River (2), Rio Grande 3 mi. upstream from the mouth of the Pecos (3), and the new Devils River bridge. This western species ranges through Trans-Pecos and east across the Edwards Plateau. It is fairly common in the study area and appears to occur mostly in the wooded areas along stream channels.

Phrynosoma cornutum. Texas horned-lizard. Fourteen specimens were collected from the following localities: Comstock, 1.5 mi. S.E. Comstock, 3.6 mi. S.E. Comstock, 4.7 mi. N.E. Comstock, 8.3 mi. N. Comstock, 12.5 mi. S.E. Comstock, 11 mi. N.W. Del Rio (3), 13 mi. N.W. Del Rio (3), 13.5 mi. N.W. Del Rio, and Langtry. This is primarily a central plains species which occurs throughout Texas with the exception of the extreme northeastern corner. It is abundant in Val Verde County, particularly in the more open areas.

Phyrynosoma modestum. Round-tailed horned-lizard.
Eleven specimens were collected from the following localities: 12 mi. W. Comstock, 13 mi. N.W. Del Rio (6), 14 mi.
N.W. Del Rio (2), mouth of the Devils River (2). This
western species reaches its eastern limits of distribution
in Val Verde County. These records represent an eastern
range extension as the species has not been recorded from
Val Verde County. The nearest previous records are from
Terrell County (Brown, 1950; Milstead, et al., 1950).
P. modestum seems quite common in the area and occupies
essentially the same habitat as P. cornutum.

Cnemidophorus gularis. Spotted whiptail. Twenty-
eight specimens were collected from the following localities: 1.4 mi. S.E. Comstock, 1.6 mi. S.E. Comstock, 1.8
mi. S.E. Comstock, 10 mi. N.W. Del Rio, 12 mi. N.W. Del
Rio (3), 13 mi. N.W. Del Rio (12), 7.8 mi. E. Langtry (2),
near mouth of the Devils River, Pecos River bridge (3),
Rio Grande 3 mi. upstream from mouth of the Pecos, and
5 mi. W. Pecos River on U.S. 90. The whiptails of western
Texas are a confusing group and I am far from convinced
that my identifications are accurate. It is possible that
more than one species is represented in this group. The
spotted whiptail ranges through much of Texas, excluding
the eastern and western edges. It is very abundant in
the study area.

Cnemidophorus tessellatus. Checkered whiptail. One
specimen was collected 12.5 mi. N.W. of Del Rio. Val Verde
County is on the extreme eastern edge of the known range
of this western species.

Cnemidophorus sp. Three specimens, two from near the
mouth of the Devils River and one from the Pecos River
bridge are not identified. They are probably referrible to
C. gularis.

Lygosoma laterale. Ground skink. Three specimens
were collected from 18 mi. N.E. Comstock and 32 mi. N.
Comstock (2). This represents the first record of this
species from Val Verde County although it has been recorded
from further west in Terrell County (Milstead, et al., 1950).
This eastern species ranges across eastern Texas to Terrell,
Tom Green, and Cooke counties.

Eumeces brevilineatus. Short-lined skink. One
specimen was collected 12 mi. N.W. of Del Rio. This species
is distributed through central Texas, east to Limestone
and Cameron counties and west to Presidio and Jeff Davis
counties, and ranges south into Mexico.

Eumeces obsoletus. Great Plains skink. One specimen was collected on the Devils River about one-half mile upstream from the mouth. It was in the process of stalking a Sceloporus merriami when collected. Snout-vent length measured 5-1/2 inches, 1/2 inch longer than the maximum reported by Conant (1958). This western species ranges through western Texas, east to Tarrant, McLennan, and Cameron counties.

Natrix erythrogaster transversa. Blotched water snake. One specimen was collected 18 mi. N.E. Comstock. This eastern species ranges across Texas to Brewster and Reeves counties. Water snakes seem to be uncommon in the study area.

Natrix rhombifera rhombifera. Diamond-backed water snake. One specimen was collected 14 mi. N.E. Del Rio. This species has been recorded previously from Val Verde County (Brown, 1950) and this is the westernmost known locality.

Thamnophis marcianus marcianus. Eastern checkered garter snake. Six specimens were collected from the following localities: Comstock, 0.9 mi. W. Comstock, 5.5 mi. N. Comstock, 6.7 mi. N. Comstock, 11.9 mi. N. Comstock, and 9.4 mi. S.E. Comstock. Val Verde County is well within the range of this western species which is found in Texas east to Tarrant, Brazos, and Matagorda counties. This seems to be about the most common of the aquatic to semi-aquatic snakes in this region, and is somewhat less dependent upon water than the others.

Thamnophis proximus ssp. Western ribbon snake. Eight specimens were collected from the following localities: 12.5 mi. N.W. Del Rio, 13 mi. N.W. Del Rio (2), 14 mi. N.W. Del Rio, and 18 mi. N.E. Comstock (4). This is an eastern species which ranges across Texas to Brewster and Reeves counties. It is fairly common around streams and permanent ponds but seems more restricted to moist habitat than is T. marcianus.

Masticophis flagellum testaceus. Western coachwhip. Twelve specimens were collected from the following localities: Comstock, 0.5 mi. W. Comstock, 6.7 mi. W. Comstock, 4.2 mi. S.E. Comstock, 6.6 mi. S.E. Comstock, 11.5 mi. S.E. Comstock, 13.2 mi. S.E. Comstock, 18.6 mi. E. Comstock, 8.9 mi. N.N.W. Comstock, 13 mi. N.W. Del Rio, 1 mi. E. Langtry, and 1.4 mi. E. Langtry. The coachwhip is a wide-ranging species which occurs across the United States. It appears to be one of the most abundant of snakes in the study area, certainly the most commonly encountered.

Masticophis taeniatus ornatus X *schotti*. Whipsnake. Two specimens were collected from 19 mi. N.E. Comstock and 21.6 mi. N. Comstock. This western species ranges through Trans-Pecos, east across the Edwards Plateau to Throckmorton, Travis, and Bexar counties, and through southern Texas to Goliad, San Patricio, and Cameron counties. These two specimens seem to be intergrades between the Balconian-Chihuahuan subspecies, *ornatus*, and the southern Texas form, *schotti*. Val Verde County is the area where the two forms would be expected to come into contact.

Opheodrys aestivus. Rough green snake. Two specimens were collected from 12.5 mi. N. Comstock and 11 mi. N.W. Del Rio. This represents the first record of the rough green snake from Val Verde County. The western limits of the known range of this eastern species are reached in Terrell County. *Opheodrys aestivus* is generally associated with moist habitat and is seemingly uncommon in Val Verde County and restricted to the canyon bottoms.

Drymarchon corais erebennus. Texas indigo snake. Two specimens were collected from 11 mi. N.W. and 13 mi. N.W. of Del Rio. This is the first record of the species from Val Verde County and the westernmost record from Texas, extending the known range some 50 miles northwest. *Drymarchon* has previously been reported from Kinney and Maverick counties (Brown, 1950). This is primarily a Mexican species which ranges through the Tamaulipan Biotic Province of Texas. Both specimens were found in heavy cover along a river terrace.

Salvadora lineata. Texas patch-nosed snake. One specimen was collected 14 mi. N.W. Del Rio. This is the first record for Val Verde County and constitutes a western range extension of about 50 miles. The species has been recorded from Kinney County (Brown, 1950). This specimen is from a critical area since it very nearly closes the gap between the ranges of *lineata* and the closely related *S. grahamiae* to the west. The relationships between *lineata* and *grahamiae* are deserving of critical study. *S. lineata* ranges through central Texas from Tarrant and Young counties, west to Val Verde and east to Calhoun counties, and south into Mexico.

Elaphe guttata emoryi. Great Plains rat snake. One specimen was collected 23.1 mi. N. of Comstock. Val Verde County is well within the range of this plains species which ranges across Texas excluding the eastern edge.

Elaphe obsoleta bairdi. Baird's rat snake. One specimen was collected 15 mi. N.W. of Del Rio. This snake has not been previously reported from Val Verde County, but is within the expected range. Baird's rat snake is a rather uncommon snake and its taxonomic relationship to *E. o. lindheimeri*, the eastern subspecies in Texas, needs evaluation. *Elaphe obsoleta* is an eastern species generally ranging west only to about the 100th meridian, but the subspecies *bairdi* extends across the Edwards Plateau to Culberson, Jeff Davis, and Brewster counties.

Elaphe subocularis. Trans-Pecos rat snake. Four specimens were collected from the following localities: 1.8 mi. E. Langtry, 8 mi. E. Langtry, 14.2 mi. W. Pecos River bridge, and 3.1 mi. E. Pecos River bridge. This species reaches its eastern range limits in Val Verde County. It is rare east of the Pecos River, and the specimen from 3.1 mi. E. of the Pecos River bridge is one of the easternmost records. All of the specimens collected were found dead on the road.

Pituophis melanoleucas sayi. Bullsnake. Two specimens were collected from 13 mi. N.W. of Del Rio and on Farm Road 1024 (Pandale Road) 25 mi. N. of U.S. Highway 90. Val Verde County is within the known range of this widely distributed species. The Pandale Road specimen measured 7.2 feet in length and contained three full-grown cottontail rabbits in its stomach. This specimen was found on the road, having been recently shot.

Lampropeltis mexicana. Mexican king snake. Three specimens were collected from 9.5 mi. E. of Comstock, 1.5 mi. W. of Comstock, and 10.6 mi. W. of Comstock. These three snakes are the most interesting specimens of the entire collection and provide some rather puzzling problems. The formerly considered distinct species *L. blairi* and *L. alterna* have recently been placed as subspecies of *L. mexicana* (Gehlbach and Baker, 1962). *L. blairi* was formerly confined to Terrell and Val Verde counties and *L. alterna* was a Trans-Pecos species. Gehlbach and Baker (1962) reported an intergrade specimen from Edwards County, east of the range of *blairi*. The specimen from 10.6 mi. west of Comstock is very near the typical *blairi* in coloration, having very broad black bands and reduced orange pigmentation. The specimen from 9.5 mi. S.E. of Comstock is similar in pattern but with very reduced black pigment and broad orange bands. The third specimen more nearly resembles *L. alterna*, being predominantly gray with very narrow bands and reduced orange pigment. The separation of these specimens into subspecies is difficult although two could be assigned to *L. m. blairi* and one to *L. m. alterna*. It is likely that all the specimens are intergrades to some extent. It is equally possible that the color and pattern variation represents polymorphism as much as hybridization. L. mexicana is a Mexican species which extends into Trans-Pecos, Texas and ranges east to Edwards and Val Verde counties. It has been considered as a rare snake east of the Pecos River, but at least five specimens were obtained in Val Verde County by various collectors during the summer of 1965.

Rhinocheilus lecontei tessellatus. Texas long-nosed snake. Four specimens were collected from the following localities: 2.0 mi. W. Comstock, 21.9 mi. N. Comstock, 16.6 mi. N. Del Rio, and 5.0 mi. E. of Langtry. Val Verde County is within the range of this western species which ranges east in Texas to Dallas, McLennan, and San Patricio counties.

Sonora episcopa episcopa. Great Plains ground snake. Four specimens were collected from the following localities: Comstock, 13 mi. N.W. Del Rio, 14 mi. N.W. Del Rio, and 1 mi. E.Langtry. This is a prairie species which ranges through Texas from Presidio County on the west to Dallas, McLennan, and Calhoun counties on the east.

Hypsiglena ochrorhyncha ochrorhyncha. Night snake. Three specimens were collected from Comstock, 12 mi. W. Comstock, and 7.5 mi. N.W. of Del Rio. A western species, the night snake extends eastward across Texas to Parker, Bosque, and Hidalgo counties. It is rather uncommon east of the Edwards Plateau.

Tantilla gracilis gracilis. Slender flat-headed snake. Two specimens were collected from 32 mi. N. of Comstock. Val Verde County represents the southwestern limits of the known range of this species in Texas. Tanzer, et al. (1966) provided the first record for the county, three specimens from Dolan Creek, 40 miles N. of Del Rio. The specimens reported here provide a slight range extension of 30 miles to the west. Tantilla gracilis is primarily a plains species, ranging across central and eastern Texas to Val Verde County. It has not been reported from Mexico, to my knowledge.

Micrurus fulvius tenere. Texas coral snake. One specimen was collected from near Comstock. Baird and Girard (1853) described Elaps tenere with the type locality given as the Rio San Pedro of the Rio Grande (=Devils River). No other specimens have been reported from Val

Verde County and the unreliability of old records has apparently led to some doubt as to the western limits of range in Texas. For instance, Conant (1958) maps the range of M. f. tenere as stopping somewhat short of Val Verde County and marks with an X, separated from the rest of the range, a western population which is apparently that reported from Terrell County (Milstead, et al., 1950). A second specimen from Val Verde County is available in the Texas Natural History Collection, The University of Texas. This specimen was collected at the mouth of the Pecos River. These records fill the gap between the Terrell County records and the rest of the species range. The coral snake is an eastern species.

Agkistrodon contortrix pictigaster. Trans-Pecos copperhead. Two specimens were collected from 27 miles N. of Comstock and from Langtry. This is the first record of the copperhead from Val Verde County and if the subspecific identification is correct extends the range of pictigaster about 50 miles to the east and across the Pecos River. Agkistrodon contortrix is an eastern species found primarily east of the 100th meridian, but extending into Trans-Pecos Texas via the Edwards Plateau and Rio Grande drainage.

Crotalus atrox. Western diamondback rattlesnake. Five specimens were collected from the following localities: 7.7 mi. W. Comstock, 17 mi. N.W. Del Rio, 1 mi. E. Langtry (2), and 9.5 mi. E. Langtry. This snake is much more common than the records indicate. It is a western species which ranges through all but extreme eastern Texas.

Crotalus lepidus lepidus. Mottled rock rattlesnake. Four specimens were collected from 1 mi. E. Langtry (2), 9.2 mi. E. Langtry, and the Rio Grande 3 mi. upstream from the mouth of the Pecos River. This Mexican species is fairly common in Val Verde County and ranges through Trans-Pecos east along the southern edge of the Edwards Plateau to Real County.

DISCUSSION

Fifty-two species of amphibians and reptiles, 10 frogs and toads, 5 turtles, 15 lizards, and 22 snakes, were recorded during this study. Of these, 11 represent the first records for Val Verde County of species whose known ranges include this area. Range extensions are reported for four species and one subspecies, Phrynosoma modestum, Drymarchon corais, Salvadora lineata, Tantilla gracilis, and Agkistrodon contortrix pictigaster. Of particular interest to herpetologists was the collection of three specimens of Lampropeltis mexicana.

In addition to the species reported above, 18 species of amphibians and reptiles have been recorded from Val Verde County (1 salamander, 2 frogs, 1 turtle, 6 lizards, and 8 snakes), giving a total of 70 species known to occur (Table 12). At least seven other species (1 salamander, 1 turtle, and 5 snakes) are unrecorded but probably occur in the county.

Species may be somewhat arbitrarily assigned zoogeographic affinities based upon their distribution. I have chosen to divide the Val Verde herpetofauna into five categories, i.e., western, eastern, Balconian, Mexican, and other. Western and eastern species are those who range primarily west and east of the 100th meridian respectively. Balconian species are those whose ranges are principally or wholly restricted to the Edwards Plateau. The Mexican species have ranges centered in Mexico and the category other includes those species which are widespread or primarily central grasslands forms. There is one endemic species, Tantilla diabola, which is known only from the type locality, Dolan Springs, 40 miles north of Del Rio.

Of the 70 species known to occur in Val Verde County, 31 (44%) are classed as western, 15 (21%) eastern, 12 (17%) other, 7 (10%) Mexican, 4 (6%) Balconian, and 1 (1%) endemic. The zoogeographic affinities of Val Verde County (based upon the herpetofauna) are predictably closest to the Chihuahuan Biotic Province to the west. There is a rather surprising influence of eastern forms when one considers the generally arid conditions that prevail. This is undoubtedly due to two factors, the presence of permanent streams, Devils River, Pecos River, and Rio Grande, which provide habitat for mesic-adapted species, and the presence of the Edwards Plateau which seems to provide a corridor for eastern species to extend far to the west (and conversely for western species to extend eastward). Most of the eastern species are aquatic, semi-aquatic, or restricted to moist environments. Their distribution in Val Verde County is, therefore, expected to be dendritic, following the stream drainages. While many of the western forms also occupy stream channel habitat, many are primarily found on the drier uplands.

Additional field work in the area will help to clarify the habitat relationships of the species and may provide additional data to help solve several knotty taxonomic problems.

TABLE 12. THE HERPETOFAUNA OF VAL VERDE COUNTY

Species recorded during this study	Species recorded in the literature	Species unrecorded but to be expected
	SALAMANDERS	
	Eurycea neotenes	_Ambystoma tigrinum_
	FROGS AND TOADS	
Scaphiopus couchi	_Syrrhophus marnocki_	
Scaphiopus hammondi	_Bufo debilis_	
Eleutherodactylus augusti		
Bufo speciosus		
Bufo punctatus		
Bufo valliceps		
Acris crepitans		
Gastrophryne olivacea		
Rana pipiens		
Rana catesbeiana		
	TURTLES	
Kinosternon flavescens	_Terrapene ornata_	_Chelydra serpentina_
Pseudemys concinna		
Pseudemys scripta		
Gopherus berlandieri		
Trionyx spinifer		

Species recorded during this study	Species recorded in the literature	Species unrecorded but to be expected

LIZARDS

Coleonyx variegatus	Sceloporus cyanogenys	
Crotaphytus collaris	Sceloporus magister	
Holbrookia texana	Cnemidophorus inornatus	
Sceloporus merriami	Cnemidophorus tigris	
Sceloporus olivaceus	Holbrookia lacerata	
Sceloporus poinsetti	Gerrhonotus liocephalus	
Sceloporus undulatus		
Urosaurus ornatus		
Phrynosoma cornutum		
Cnemidophorus gularis		
Cnemidophorus tessellatus		
Lygosoma laterale		
Eumeces brevilineatus		
Eumeces obsoletus		

SNAKES

Natrix erythrogaster	Leptotyphlops dulcis	Arizona elegans
Natrix rhombifera	Leptotyphlops humilis	Lampropeltis doliata
Thamnophis marcianus	Diadophis punctatus	Lampropeltis getulus
Thamnophis proximus	Heterodon nasicus	Tantilla nigriceps

Table 12 (cont'd)   207

Species recorded during this study	Species recorded in the literature	Species unrecorded but to be expected
Masticophis flagellum	Tantilla diabola	Sistrurus catenatus
Masticophis taeniatus	Thamnophis cyrtopsis	
Opheodrys aestivus	Agkistrodon piscivorus	
Drymarchon corais	Crotalus molossus	
Salvadora lineata		
Elaphe guttata		
Elaphe obsoleta		
Elaphe subocularis		
Pituophis melanoleucas		
Lampropeltis mexicana		
Rhinocheilus lecontei		
Sonora episcopa		
Hypsiglena ochrorhyncha		
Tantilla gracilis		
Micrurus fulvius		
Agkistrodon contortrix		
Crotalus atrox		
Crotalus lepidus		

VERTEBRATE PALEOFAUNA OF AMISTAD RESERVOIR

Gerald G. Raun

METHODS

Seven archeological sites were sampled for vertebrate faunal remains. Three of these are not considered here; Devils Rockshelter and Mosquito Cave contained too few bones to be of any significance and Castle Canyon provenience was unavailable until too late to allow analysis. Castle Canyon should be analyzed in the future, although only one of the several subsections yielded any bone. The following sites were analyzed in some detail: Eagle Cave (41 VV 167), Zopilote Cave (41 VV 216), Coontail Spin (41 VV 82), and Devil's Mouth (41 VV 188). Of these, Eagle Cave and Coontail Spin were the most productive.

None of these sites were collected specifically for faunal remains. Bone material was collected during routine archeological excavation and as a result the vertebrate remains may not be an accurate sample of the preserved fauna. Small rodent material, which escaped most screening techniques used in the field, is particularly scarce. Where one might expect a reasonable amount of loose rodent teeth, there are none or only a very few. The relative frequency of larger bone, such as deer or rabbit, is probably exaggerated in most of the samples because of collecting technique, differential preservation, and ease of identification.

Identifications were based primarily upon cranial elements. Postcranial bones of deer and rabbit were identified when complete. No effort was made to identify postcranial bones of rodents. This can probably be done, but not in the time available for this preliminary study. This would require an extensive reference collection, and considerable time. It should be considered in future projects.

In general, identification was made only at the generic level. Species names are attached only if the bone material was diagnostic and no names have been assigned on the basis of present geographic range. This is extremely dangerous practice, particularly when dealing with older deposits.

Frequency of occurrence is expressed as a percentage of total identifiable bone in a given strata (Figures 30-32). Scrap and unidentifiable bones were counted but do not enter into any of the discussion or calculations.

209

THE COLLECTIONS

Coontail Spin Site

The Coontail Spin Site consisted of two discrete units which were analyzed separately (see report in Archeological Background). Area A contained the greatest concentration of bone and provided the most diversified faunal list of all sites examined (Table 13). Area B was much less productive. Whether this reduction resulted from collecting techniques or from actual differences in deposition or preservation is open to question.

Only one of the species identified from Area A is not a member of the modern fauna. The muskrat (_Ondatra_ cf. _zibethecus_), represented by a single lower jaw, is now found in extreme southeastern Texas and in the Big Bend region of western Texas. All the remaining forms are to be expected in the area today with the possible exception of _Taxidea_ cf. _taxus_. Val Verde County is within the range of the badger but ecological conditions do not now seem favorable for badger.

Some of the canid material is rather interesting. Apparently _Canis_ cf. _latrans_ and _Urocyon_ cf. _cinereoargenteus_ were present and also a rather short-faced canid which may have been a domesticated dog. A few fragmentary mandibles seem too short to be assigned to coyote. However, separation of dog cranial elements from those of coyote is a difficult task at best and the material available is far from adequate. Almost half of the bones identified as _Canis_ sp. were found in the 11 to 12 foot level where there were a number of cranial bones present. Certainly more than a single individual was represented, but no guess can be made as to the number. Other canid bones were scattered through the site.

Coontail Spin (both A and B) contained a greater relative proportion of fish remains than any other site with the possible exception of Devil's Mouth. This is particularly true of the upper levels (Figure 30) and the relative frequency of fish bone decreases with depth. This could be an accident of collecting or preservation, but I feel that the trend is probably real. Fish bone is generally more resistant than is rabbit bone, for example, and the collecting technique presumably did not vary from one zone to another.

210

Rabbit was an important element and in combination with fish provided in excess of 50% of the bone in all zones except the surface and the 11 to 12 foot layer. Very few bones were recovered from the 11 to 12 foot layer and an unusually high percentage of Canis sp. tends to overshadow the rabbit and fish material. The relative frequency of rabbit increases with depth (Figure 30) and in the lower zones replaces fish as the most important element. Both cottontails (Sylvilagus sp.) and jack-rabbits (Lepus cf. californicus) are present with the former appearing in much greater frequency. This preponderance of Sylvilagus appears directly opposite to the present situation in Val Verde County. Lepus is now by far the more abundant and cottontails seem scarce.

Deer bones were present in all zones and provided a significant part of each sample. The percent frequency varied little from top to bottom in the site.

The fourth major item was turtle bone, usually carapace or plastron fragments. Soft-shelled turtle (Trionyx sp.) was represented most frequently. The only other identification that could be made was Pseudemys sp.. The latter was present in very small amount, much less abundantly than Trionyx. This I find somewhat surprising since Pseudemys is usually easier to catch than Trionyx. If it is correct to assume that the abundance of bone in the site reflects the abundance of material originally present and is not related to differential preservation, then it seems apparent that the inhabitants had developed rather sophisticated devices for obtaining turtles. Soft-shelled turtles seldom come out on the bank where they are accessible to hand collection and they are, in addition, extremely wary. It is probable that capture of soft-shelled turtles in any number would require some type of trap. Turtle bone is more abundant in the upper zones than in the lower, generally following the pattern of fish remains.

The remaining forms occur rather sporadically through the site except for woodrats (Neotoma sp.) and cottonrats (Sigmodon cf. hispidus), both of which are present in most samples but in low numbers.

The presence of porcupine (Erethizon cf. dorsatum) is rather interesting, but not as significant as one might think after looking at the more recent range maps (Hall and Kelson, 1959: 782, for instance). Val Verde County is

FIGURE 30. Relative frequency of identified faunal
            remains, Coontail Spin Site.

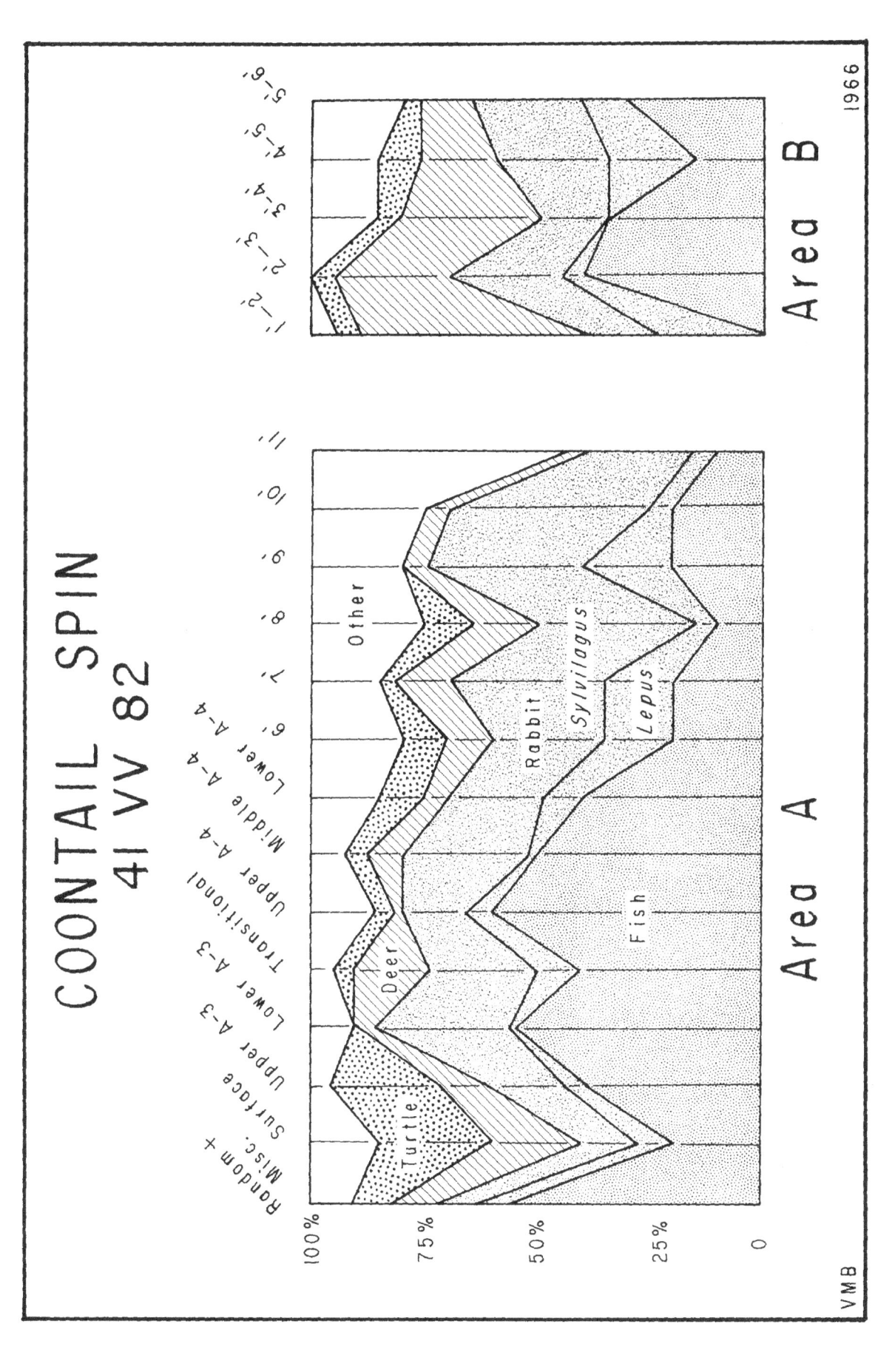

COONTAIL SPIN
41 VV 82

Area A

Area B

1966

VMB

more than 100 miles east of the mapped range, but as Milstead and Tinkle (1959) have shown, porcupines are apt to appear far from their normal habitat. L. J. Eck, while engaged in field research for this study, observed a freshly-killed specimen on the highway about one and a half miles north of Comstock in the summer of 1965, showing that they do occur, on occasion, in Val Verde County. The area is about as atypical of porcupine habitat as one could imagine. Beaver (Castor cf. canadensis) was also identified in the deposit, and while no evidence of recent beaver activity was noted, they are to be expected along the rivers. The scarcity of modern beaver is more probably related to human activity than to ecological change.

Area B was much less productive of faunal remains but does exhibit some interesting differences. Deer was present in greater proportion than in Area A and in the upper zones is as important or more important than rabbit.

Bone was well distributed throughout Area A with the greatest concentration being found in the upper portion of Zone A-4. Bone concentration fell off sharply below the 11 to 12 foot level. Only a few scraps were found in the 12 to 13, 13 to 14, and 14 to 15 foot levels, and in lower Zone A-4. Above the 6 foot level of Area B, bone was evenly distributed but was scarce below.

Eagle Cave

The fauna recovered from Eagle Cave is only slightly less diverse than that of Coontail Spin and in some respects is more interesting. Four species were present which were not found at Coontail Spin. Two of these, Mexican ground squirrel (Citellus mexicanus) and skunk (Mephitis sp.), are of little significance while the pocket gopher (Geomys sp.) and bison (Bison cf. bison) are of more interest.

Geomys personatus is reported to occur in eastern Val Verde County, along the Rio Grande at Del Rio (Davis, 1940: 31). Only two gopher jaws, both in Stratum V, were found, but their presence is indicative of fairly deep sandy soils. Such conditions are now found only along the river terraces and no evidence of gopher activity has been found in the study area.

213

Bison bones were recovered from Strata I, IIa, and III. These were badly broken and only a few (14) were found. The only other site containing Bison was Bonfire Shelter, which is the subject of a separate report (see report herein by Lorrain). Since Bison is essentially confined to Bonfire Shelter, one might assume that they were present in Val Verde County only on occasion and were not a common or usual element of the fauna at the time of deposition. Most of the bone appears to have been from young animals.

Deer bone was the most important single element in Stratum I and rabbit bone was not found. Below this zone, however, the relative frequency of deer bone decreased and rabbit increased (Figure 31). In Strata IV and V deer bone was essentially absent. Whether this decrease in deer and corresponding increase in rabbit is related to the relative abundance of the animals is purely speculative. It is somewhat difficult to envision the possibility that deer were less abundant at the time these lower units were laid down than at the present or during the time of deposition of Stratum I. Perhaps this variation in the site is a result of hunter efficiency. That is, the earlier hunters were less able to kill deer.

Here, as in Coontail Spin, Sylvilagus remains were identified in greater proportion than were those of Lepus. In fact, the disproportion is even more distinct. As stated above, the relative abundance of the bone is the reverse of what would be expected from present conditions.

In Eagle Cave, as in Coontail Spin, soft-shelled turtle remains far outnumber those of other turtles.

Zopilote Cave

A fair faunal list was compiled from this site although the number of bones recovered was quite small. This site was dug in half-foot levels and all levels above 2 feet were essentially sterile of bone. A single identifiable bone was found in each of the 6-inch layers from the surface to 2 feet below the surface. Greater concentrations were recovered from 2 to 5 feet but at no time was there sufficient bone to provide any realistic faunal pattern. Below the 5 foot level only two rabbit bones were found.

FIGURE 31. Relative frequency of identified faunal remains, Eagle and Zopilote Caves.

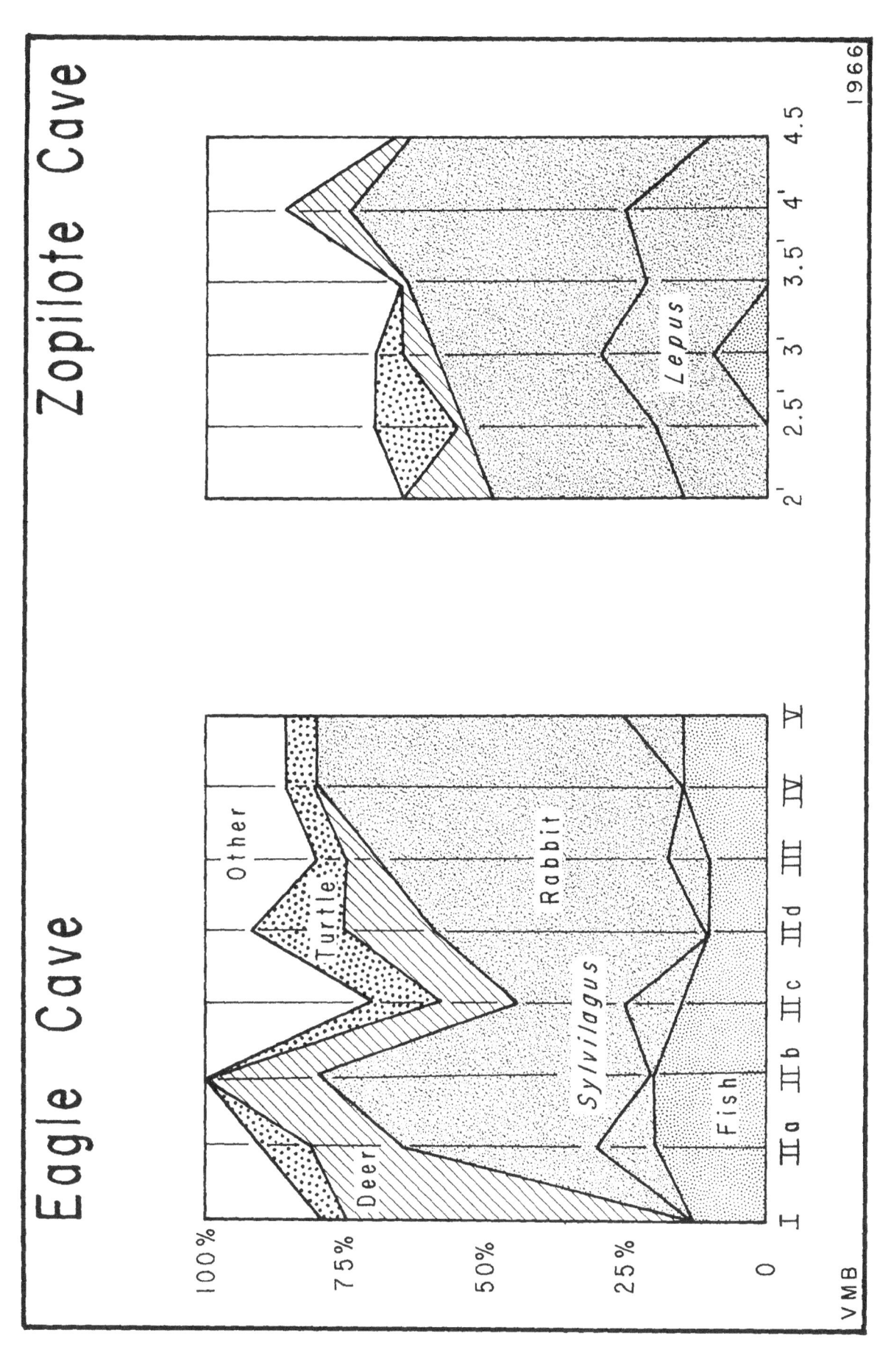

FIGURE 32.   Relative frequency of identified faunal
remains, Devil's Mouth Site.

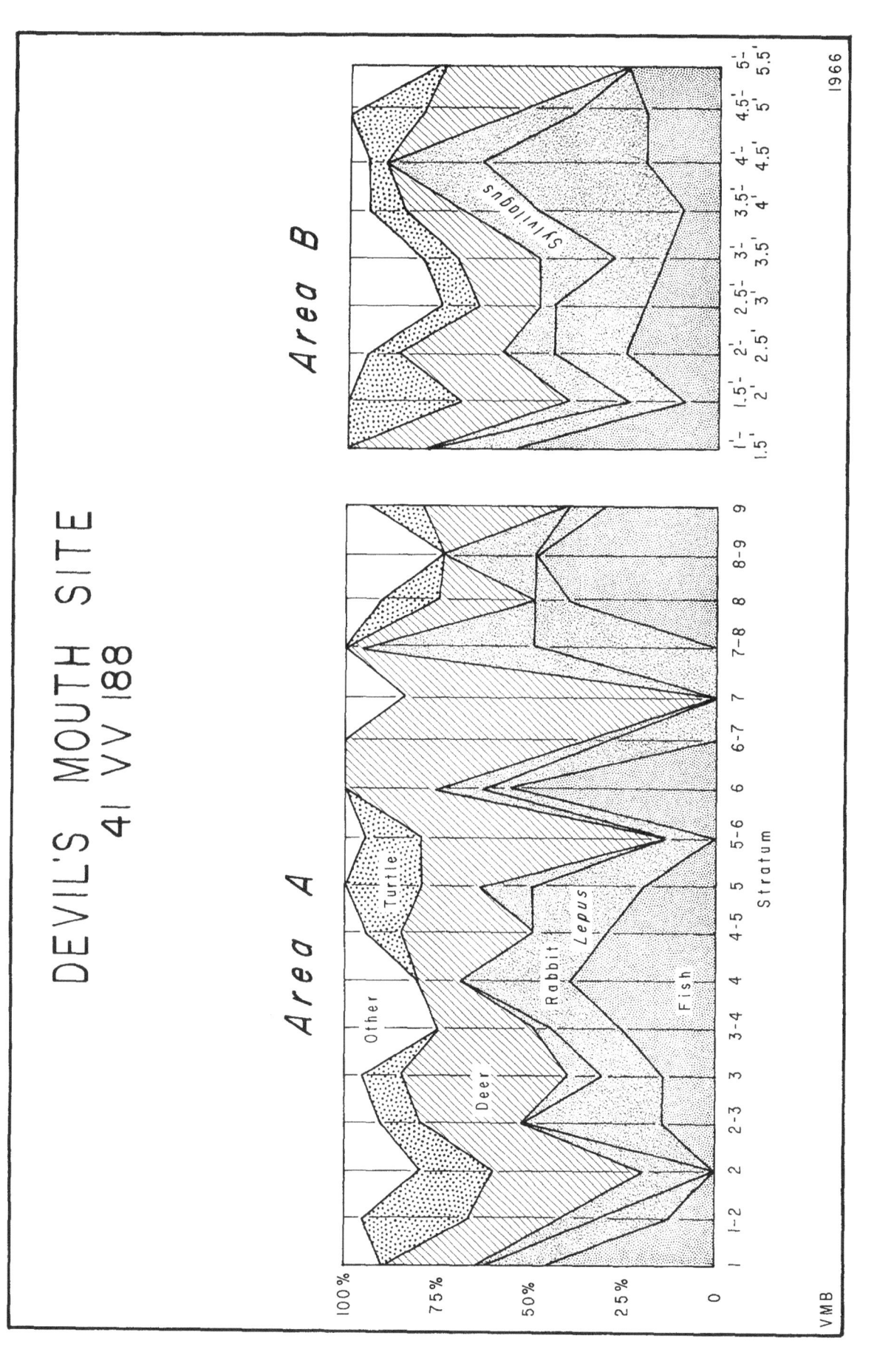

DEVIL'S MOUTH SITE
41 VV 188

Area A

Area B

VMB

1966

Rabbit was the most abundant in this site (Figure 31) and fish, turtle, and deer were relatively unimportant. Again, _Sylvilagus_ was more abundant than _Lepus_, but the distinction was less than seen previously. Fish bone, an important element in all other sites, was practically absent from Zopilote Cave. This is also true of turtle remains, and interestingly, all the turtle material appears to be from terrestrial forms. This is the only site from which _Trionyx_ is absent. Apparently, aquatic food items were not brought to this site in any abundance.

Devil's Mouth Site

Devil's Mouth is a terrace site as opposed to the others which are rockshelters. As such, one might expect to find species present which were not being utilized as human food items (assuming that the remains from the caves were primarily the result of human activity). Unfortunately, this expectation was not realized at this site. With the exception of _Thomomys_ and _Cratogeomys_ all other members of the preserved Devil's Mouth fauna were present in other sites.

Two separate units, A and B, were excavated and the total number of identifiable remains was low. The derived faunal list is shorter than those of other sites, but in some respects was more interesting. This interest stems primarily from the presence of three geomyid rodents, _Geomys_ sp., _Thomomys_ sp., and _Cratogeomys_ cf. _castenops_. The latter two were not found elsewhere.

Thomomys bottae has been recorded from Val Verde County at the following localities: Comstock; Devils River, 13 miles below Juno; and Samuels, 17 miles west of Langtry (Baker, 1953: 505). Thus, the present range of _Thomomys bottae_ would appear to encompass most, if not all, of the study area. With the exception of an isolated population reported from 35 miles east of Rocksprings, in Real (?) County (Goldman, 1936: 119), the Val Verde County records mark the eastern limits of the known range in Texas. If _Thomomys_ does occur in the study area at this time, it must be confined to restricted and widely scattered colonies. No evidence of its presence was discovered during this study. _Thomomys_ seems to prefer thin, rocky soils (where neither _Geomys_ or _Cratogeomys_ could exit), but even thin soils are scarce in most of the study area. One _Thomomys_ sp. jaw was found in mixed strata 4 and 5 of Area A, and one in mixed strata 8 and 9. It was not recovered from Area B. The other two gophers were found in Area A but not Area B.

Cratogeomys is the largest of the three gophers and
requires a rather thick soil layer in which to burrow.
The known range of Cratogeomys castenops includes the study
area. Val Verde County records are from Juno; Langtry;
and Samuels, 17 miles west of Langtry (Nelson and Goldman,
1934: 139). No evidence of current Cratogeomys activity
in the Amistad area was found. Like Thomomys, it must
occur in very restricted, scattered colonies if it is still
present. Suitable habitat is seemingly available along the
river terraces which would support both Geomys and Cratogeomys, but neither could colonize the vast majority of
the upland habitat. One Cratogeomys jaw was recovered from
the 2.5 to 3.0 foot layer in Area B, and one was present
in the 5.0 to 5.5 foot level, Area B.

A single Geomys jaw was found in the 3.0 to 3.5 foot
level. It is possible that all of the geomyid remains from
Devil's Mouth could have been brought in by avian predators.
The cruising radius of most of these predators is restricted
enough to indicate gopher activity in the vicinity of the
site during deposition.

The occurrence of gophers in the site and their
apparent absence now could be taken to indicate that some
ecological disturbance (removal of top soil) may have occurred. Such an interpretation should be made with some
caution, however, since the evidence is far from conclusive.

Patterns of relative abundance of skeletal remains
from Devil's Mouth (Figure 32) are of limited interpretive
value because of the low concentration of bone. This is
the only site in which the relative frequency of jackrabbit bone exceeded that of Sylvilagus. It is also the
only site in which observed frequency of deer bone
approached that of rabbit.

Fluctuations in percentage frequency in Area A are too
great to permit the observance of any pattern. Area B is
only slightly more reliable. With the exception of the
jackrabbit-cottontail ratio, the general trends seen in
Area B are similar to those from other sites (Figs. 30,31).

ECOLOGICAL IMPLICATIONS

One glaring discrepancy in the data derived from these
sites is the almost complete absence of smaller rodents.
Sigmodon, Citellus spilosoma, and C. mexicanus are the
smallest rodents represented. Jaws of this size are about

the smallest that one could expect to trap by standard
archeological screening techniques. Since smaller screens
were not employed by the field crews excavating these sites,
sampling error probably plays a major role in this absence
of smaller materials. Test samples of dirt were taken from
Bonefire Shelter and fine-screened. A very few rodent
teeth were recovered (mostly Sigmodon and Neotoma) because
the material was badly charred and most of the small bone
and teeth were probably destroyed by the fire. Extensive
fine-screening and washing of material from future sites
should prove helpful in overcoming some of the problems
encountered to date.

If the accumulation of bone was solely the result of
human activity (i.e., representing food items) one would
expect the larger forms to be more abundant in the sample
than the smaller species. Superficially, this would appear
to be the situation in the sites examined here. Unfortunately, the sampling error prevents any such conclusion.
It is only possible that most of the represented animals
were utilized by humans as food and that a few were in the
shelters "voluntarily." It is possible that all represent
food items.

All species listed in Table 13 are to be expected in
the study area today with the exception of Geomys sp.,
Ondatra cf. zibithecus, and Bison cf. bison. It is also
likely that Thomomys sp., Cratogeomys cf. castenops, and
Taxidea taxus are not now found in the vicinity of the
sites. This is insufficient data from which to draw ecological implications, although there is some suggestion
that there has been erosion of top soil and perhaps some
drying of the area during the past 4,000-5,000 years (see
Appendix A of the report on pollen from the Devil's Mouth
Site). It is also possible that much of this change, if
indeed it did occur, may be of very recent origin related
to poor range management practices.

The observed difference in the relative frequency of
cottontails and jackrabbits in the paleofauna as compared
to the modern fauna could indicate some ecological change
since the time of deposition. Although these rabbits
occupy similar habitat over a rather wide area, jackrabbits
seem to be favored by more arid conditions and are admira-
bly adapted to grasslands. Sylvilagus generally prefers
more cover and suitable habitat is largely restricted to
areas along the river terraces. Cover is virtually absent
on the uplands which are usually overgrazed to the point

TABLE 13.

FAUNAL LISTS DERIVED FROM VERTEBRATE MATERIAL FOUND
IN FOUR ARCHEOLOGICAL SITES

Species	Eagle Cave	Coontail Spin A	Coontail Spin B	Zopilote Cave	Devil's Mouth A	Devil's Mouth B
Fish (unidentified)	X	X	X	X	X	X
Turtle (unidentified)	X	X	X	X	X	X
Pseudemys sp.	X	X	-	-	X	X
Trionyx sp.	X	X	X	-	X	X
Bird (unidentified)	X	X	X	X	X	X
Sylvilagus sp.	X	X	X	X	X	X
Lepus cf. californicus	X	X	X	X	X	X
Neotoma sp.	X	X	X	X	X	X
Sigmodon cf. hispidus	X	X	-	X	X	X
Citellus sp.	-	X	X	X	-	X
C. variegatus	X	X	-	-	-	-
C. spilosoma	X	X	-	-	-	-
C. mexicanus	X	-	-	-	-	-
Geomys sp.	X	-	-	-	-	X
Thomomys sp.	-	-	-	-	X	-
Cratogeomys cf. castenops	-	-	-	-	-	X
Ondatra cf. zibethecus	-	X	-	-	-	-

TABLE 13 (cont'd)

Species	Eagle Cave	Coontail Spin A	B	Zopilote Cave	Devil's Mouth A	B
Erethizon cf. dorsatum	X	X	-	-	-	-
Castor cf. canadensis	-	X	-	X	X	-
Canis sp.	X	X	-	X	X	-
Urocyon cf. cinereoargenteus	-	X	-	X	X	X
Canid (unidentified)	-	X	X	-	-	-
Felid (unidentified)	X	X	X	-	-	-
Procyon cf. lotor	-	X	X	X	-	-
Bassariscus cf. astutus	-	X	X	X	-	-
Taxidea cf. taxus	X	X	X	-	-	-
Spilogale cf. gracilis	-	X	-	X	-	-
Mephitis sp.	X	-	-	-	-	-
Mustela cf. frenata	-	X	-	-	-	-
Odocoileus sp.	X	X	X	X	X	X

of denudation. It would be unwise to postulate a general
ecological shift from a more wooded habitat to sparse
grassland solely on the basis of frequency of rabbits for
at least two reasons. Cottontails are probably much easier
to obtain by snaring or hunting. Even if the Indians had
no particular preference for one over the other, they would
probably kill more cottontails than jackrabbits. Also,
some of the animal remains may well have been brought in by
predators, particularly owls. An adult jackrabbit may be
approaching the maximum size limit of prey for most owls
(except the great-horned owl) and might be expected to
appear in owl pellet accumulations less frequently than
the smaller cottontails.

BONFIRE SHELTER FAUNA

Dessamae Lorrain

The following faunal list is based largely upon the identifications made by Mr. Reuben Frank (Dibble, 1965, Appendix 1) although I have added the bison species and re-examined all of the bones with the help of Mr. Bob Slaughter of the Shuler Museum of Paleontology of Southern Methodist University. No attempt was made to determine the number of individuals of any one species; only their presence or absence is recorded (Table 14). In general, if more different species were present in a given stratum then a larger number of individuals of the common species (jackrabbit, rabbit, pocket mouse, and wood rat) were present. However, in considering the ecological significance of the fauna, it is necessary to take into account the nature of the strata, the manner of formation of the bison bone beds and the number of individuals of the two bison species.

Bone Bed 1 contained scattered and badly fragmented remains of the various animals in a matrix of small to large limestone spalls. Bone Bed 2 consisted largely of the butchered remains of now-extinct species of bison in a deposit of sandy silt and limestone spalls. These animals had been driven over the canyon rim, then slaughtered and butchered in the rockshelter by Paleo-Indian (Period I) hunters. At least three separate drives were represented in Bed 2, resulting in the death of an estimated total of 120 individual animals (Dibble, 1965, Appendix 2). Bone Bed 3 (Period V) was a thick deposit of burned, cut, and broken bones of Bison bison. Some silt and occasional limestone spalls were present among the bones but formed only a small part of the total volume of the bed. It is estimated that at least 800 individuals are represented. It is unknown whether one large herd or several herds are included in this bone accumulation. It does seem clear, however, that only a relatively short span of time is represented. The Fiber Layer (Period VI) was a thin cultural deposit composed primarily of vegetal material including lecheguilla leaves, prickly pear pads, twigs, and various hulls and seeds. The only other cultural stratum at the site was the Intermediate Horizon (Period II?), a thin scattering of occupational debris within Zone 2. Zones 1, 2, and 3 were culturally sterile soil zones between the bone beds. Figure 33 shows the ages of the zones and bone beds as determined by the radiocarbon method.

FIGURE 33.   Idealized profile and radiocarbon dates,
             Bonfire Shelter.

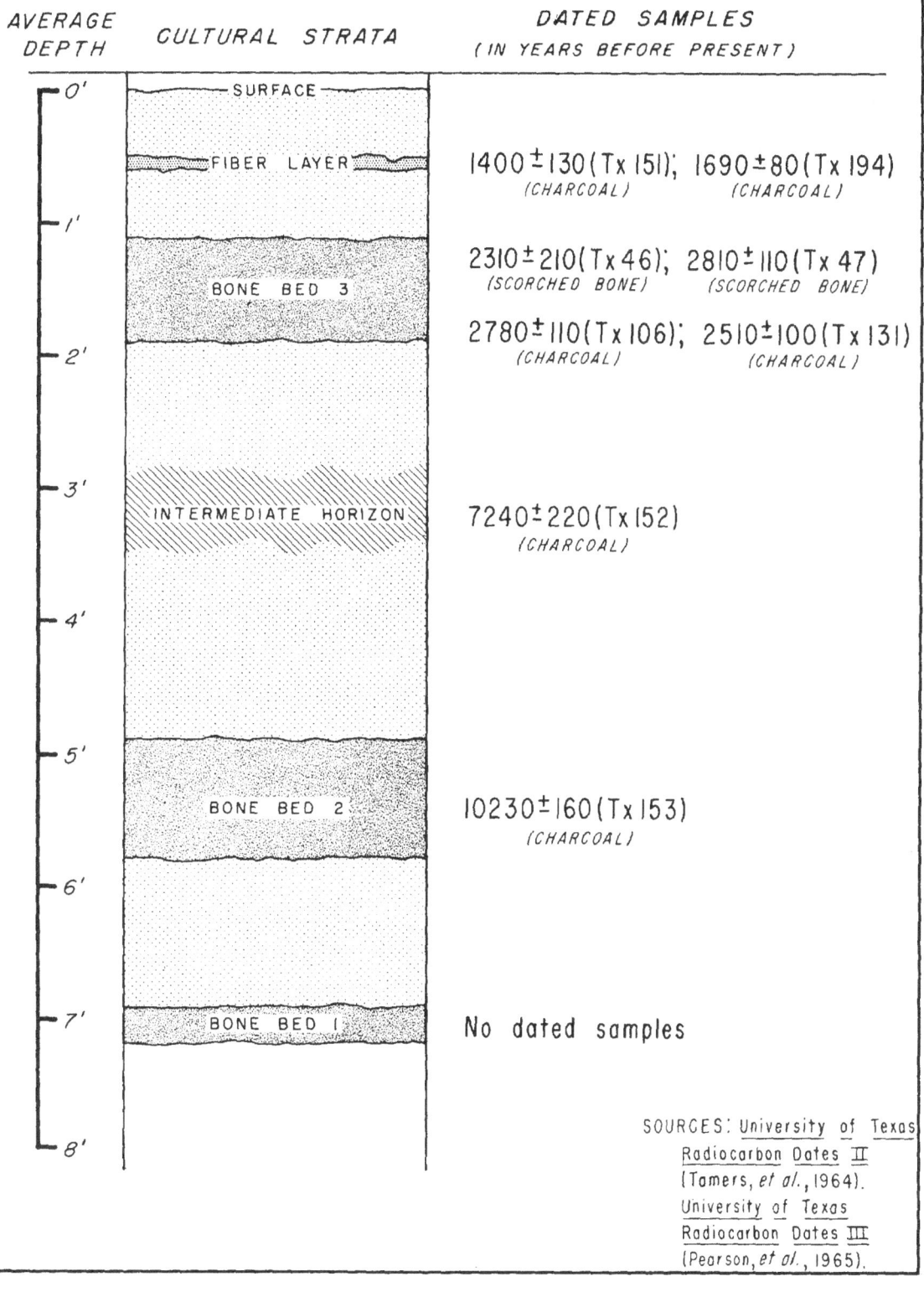

IDEALIZED PROFILE OF CULTURAL DEPOSITS WITH ASSOCIATED RADIOCARBON DATES

AVERAGE DEPTH   CULTURAL STRATA   DATED SAMPLES (IN YEARS BEFORE PRESENT)

0' — SURFACE

FIBER LAYER   1400±130(Tx 151); 1690±80(Tx 194)
(CHARCOAL)   (CHARCOAL)

1'

BONE BED 3   2310±210(Tx 46); 2810±110(Tx 47)
(SCORCHED BONE)   (SCORCHED BONE)

2780±110(Tx 106); 2510±100(Tx 131)
(CHARCOAL)   (CHARCOAL)

2'

3'   INTERMEDIATE HORIZON   7240±220(Tx 152)
(CHARCOAL)

4'

5'

BONE BED 2   10230±160(Tx 153)
(CHARCOAL)

6'

7'   BONE BED 1   No dated samples

8'

SOURCES: University of Texas
Radiocarbon Dates II
(Tamers, et al., 1964).
University of Texas
Radiocarbon Dates III
(Pearson, et al., 1965).

The presence of now-extinct large herbivores--horse, bison, camel, elephant--in Bone Bed 1 indicates a more favorable climate than the present day semi-arid conditions in the vicinity of Bonfire Shelter. The vegetation of the region could not now support such animals. It is assumed with considerable confidence that Zone 1 and Bone Bed 1 date before 10,000 B.P. since they underlay Bone Bed 2 from which a C-14 date of 10,230 $\pm$ 160 B.P. was obtained. The date and the presence of extinct animals indicate a late Pleistocene age for the deposit. Other faunal studies in Texas (Slaughter and Hoover, 1963; Patton, 1963) and in eastern New Mexico (Slaughter, n.d.; Wendorf, 1961) for the same general time period have resulted in an hypothesized climate which was more moist, with a more extensive cloud cover, resulting in cooler summers and warmer winters than today. The increase in precipitation need not have been large if, as postulated, the rate of evaporation was reduced by the cloud cover. Even now, in favorable years and in areas which are not badly overgrazed, a good grass cover appears.

After the formation of Bone Bed 1 accumulated, the composition of the deposit changed markedly from primarily limestone spalls with only a small amount of silt to a deposit comprised largely of silt with only a few limestone spalls. At the time that Bone Bed 2 was deposited an equitable climate similar to that of Bone Bed 1 times might have prevailed although the sharp decrease in the rate of spalling from the walls and ceiling of the shelter indicates somewhat dryer and warmer climate. The spalling is apparently due to frost action--an action drastically reduced when the climate is warm and arid. The Bone Bed 2 bison were very probably native to the area, being regularly pursued by Paleo-Indian hunters. There must have been more forage in the area than at present, and probably a more moist climate.

No extinct species were found in any stratum above Bone Bed 2. The very few animals bones in Zone 2 compared with the relatively large number in Zone 3 would seem to indicate an unfavorable climatic interval during the deposition of Zone 2. However, this may be simply a result of the collecting since more of Zone 3 was sifted through a fine screen resulting in the collection of more small animal bones. Only two of the animals found in the strata above Bone Bed 2 (Ophisauras sp. and Perognathus flavus) are unreported for the Amistad area today. The glass lizard is common in East Texas and has been found as far west as Kerr County (Blair, 1950) which is only

120 miles east of Langtry. It seems possible that its range has contracted in the last thousand years. The silky pocket mouse has been reported just west of the Big Bend region about 200 miles west of Langtry. It lives there in an environment very much like the conditions in the Bonfire Shelter area. Both places are included in the Chihuahuan Biotic Province (Blair, 1950). It is entirely possible that the animal would be found around Langtry if a thorough collection of small mammals were made. Neither the glass lizard nor the silky pocket mouse indicates a climate unlike the present one.

The large number of bison in Bone Bed 3 could be interpreted as an indication of a more favorable climate if considered alone. However, when the facts that Bone Bed 3 and Bonfire Shelter as a whole are not typical of the archeological sites in the area are considered, a different conclusion must be reached. Bison bones are extremely scarce in other sites--deer and small mammals predominate--so it is reasonable to assume that bison were not common any time after Period I. The Bone Bed 3 bison were probably south of their normal range, driven there by an unusually hard winter or following the grass south during a particularly favorable year or years. No ecological significance can be attached to their presence here over such a short period of time.

The faunal sequence at Bonfire Shelter indicates a more moist, equitable climate before 10,000 B.P. with a gradual change to semi-arid conditions similar to the present. No major fluctuations are indicated after the deposition of Bone Bed 2, although occasional short periods of more mesic conditions probably occurred intermittently as they do today. Possibly a xeric interval is indicated during the deposition of Zone 2 (10,000 to 2,500 B.P.) unless the paucity of animal bones from this zone is merely a reflection of the collecting methods. This period would cover the span of the Altithermal which has been described as hot and dry (Antevs, 1955). Zone 2 could, at least in part, be a manifestation of this arid interval.

TABLE 14. BONFIRE SHELTER FAUNAL LIST

Classification	Zone 1	Bone Bed 1	Bone Bed 2	Zone 2	Bone Bed 3	Zone 3	Fiber Layer	Zone 3b & Fiber Layer
Class Reptilia								
Order Chelonia								
Terrapene sp. (land tortoise)	-	-	-	-	-	x	-	-
Order Sauria								
Ophisaurus sp. (glass lizard)	-	-	-	-	-	x	-	-
Class Mammalia								
Order Lagomorpha								
Lepus sp. (jackrabbit)	-	x	-	-	x	x	-	-
Sylvilagus sp. (rabbit)	-	-	-	x	x	x	x	x
Order Rodentia								
Citellus mexicanus (mexican ground squirrel)	-	-	-	-	-	x	-	-
Citellus variegatus (rock squirrel)	-	-	-	-	-	-	x	-
Geomys sp. (eastern pocket gopher)	-	-	-	x	-	x	-	-
Thomomys sp. (smooth toothed pocket gopher)	-	-	-	-	-	x	-	-

TABLE 14 (cont'd)

Classification	Zone 1	Bone Bed 1	Bone Bed 2	Zone 2	Bone Bed 3	Zone 3	Fiber Layer	Zone 3b & Fiber Layer
Order Rodentia (cont'd.)								
Cratogeomys sp. (yellow pocket gopher)			x					
Perognathus sp. (pocket mouse)	x		x	x		x		
Perognathus cf. flavus (silky pocket mouse)						x		
Reithrodontomys sp. (harvest mouse)						x		
Peromyscus sp. (white-footed mouse)			x					
Peromyscus cf. maniculatus (deer mouse)			x			x		
Peromyscus cf. leucopus (white-footed mouse)						x		
Onychomys cf. leucogaster (grasshopper mouse)						x		
Sigmodon hispidus (hispid cotton rat)			x			x		
Sigmodon sp. (cotton rat)						x		
Neotoma sp. (wood rat)	x		x			x		
Neotoma cf. micropus (southern plains wood rat)						x		

TABLE 14 (cont'd)

Classification	Zone 1	Bone Bed 1	Bone Bed 2	Zone 2	Bone Bed 3	Zone 3	Fiber Layer	Zone 3b & Fiber Layer
Order Carnivora								
Urocyon cinereoargenteus (gray fox)	-	-	-	-	-	-	-	x
Order Proboscidea								
Elephas sp.* (elephant; includes 1 mammoth tooth)	-	x	-	-	-	-	-	-
Order Perissodactyla								
Equus sp.* (horse)	-	x	x	-	-	-	-	-
Order Artiodactyla								
Bison sp.* (bison)	-	x	x	-	-	-	-	x
Bison bison (bison)	-	-	-	-	x	x	x	-
Camelops sp.* (camel)	-	x	-	-	-	-	-	-
Unidentified								
fish	-	-	-	-	-	x	x	-
snake	-	-	x	-	-	x	-	-
bat	-	-	-	-	-	x	-	-
deer	-	-	-	-	x	-	x	-
bird	x	-	-	-	x	x	-	-
rabbit	x	-	-	-	-	x	-	-

*extinct species

REPORT ON MOLLUSK SHELLS RECOVERED FROM FOUR

ARCHEOLOGICAL SITES IN THE AMISTAD RESERVOIR

E. P. Cheatum

The majority of the shells from excavated sites sent to me for identification were shells of the larger species of gastropods and unionid bivalves. It is quite natural that inexperienced collectors would gather the larger shells and overlook shells of the smaller species. Unfortunately the shells of the smaller species, at least many of them, are better indicators of climatic changes than are shells of the larger species. Since I felt that many of the smaller and more significant species had been overlooked, I made a trip in June to the Amistad area to take additional samples at the excavation sites and also to collect recent shells in that area. Subsequently to this trip I sent my assistant, Cuyler Leonard, and another one of our graduate students, John Kankrlik, to make more intensive and extended collections of shells in that area. The results obtained by these later collections were quite revealing, and doubtless more intensive collecting at the excavation sites by experienced collectors may uncover fossil shells that, from the standpoint of species and quantity, may cast a significant aspect on ecological conditions as they existed in that area when the snails were still living.

The tabulations which follow include identified specimens collected both by the archeologists and by me and my assistants, with the latter collections making up the bulk of the identifications. Although molluscan remains were collected from eight Amistad sites, only four, two terrace (Devil's Mouth and Devils Rockshelter) and two rockshelter (Eagle Cave and Bonfire Shelter) sites are incorporated into the present study. Three of the remaining sites (Centipede Cave, Cammack Sotol Pit, and Castle Canyon) yielded insignificant lists of shells, while the fourth (41 VV 263) contained a rather impressive array of species, but the archeological investigations were too limited to permit placing the shells into a meaningful context.

227

ALPHABETICAL LISTING OF IDENTIFIED MOLLUSK SHELLS

Amblema sp.

Archatinid sp.

Bulimulus alternatus

Bulimulus dealbatus

Bulimulus schiedeanus

Bulimulus sp.

Catinella vermeta

Discus cronkhitei

Durangonella sp.

Gastrocopta contracta

Gastrocopta cristata

Gastrocopta pellucida hordeacella

Gastrocopta procera

Gastrocopta tappiniana

Gastrocopta sp.

Gyraulus parvus

Hawaiia minuscula

Helicodiscus parallelus

Helicodiscus singleyanus

Helisoma anceps

Helisoma trivolvis

Helisoma sp.

Alphabetical Listing, cont'd.

Laevapex fuscus (Ferrissia rivularis)

Lamellaxis sp.

Lampsilis sp.

Physa anatina

Physa gyrina

Pisidium nitidum

Planorbis sp.

Polygyra dorfeuilliana

Polygyra texasiana

Polygyra sp.

Proptera sp.

Punctum vitreum

Pupoides albilabris

Sphaerium striatinum

Succinea avara

Succinea sp.

Tropicorbis obstructus

I. MOLLUSK SHELLS FROM EAGLE CAVE (41 VV 167)

Amblema sp.

Bulimulus alternatus

Bulimulus dealbatus

Intrasite Distribution:

    Stratum IIa:

        Bulimulus alternatus

    Stratum IIc:

        Bulimulus dealbatus

    Stratum IId:

        Amblema sp.

        Bulimulus alternatus

    Stratum III:

        Bulimulus alternatus

    Stratum V:

        Bulimulus alternatus

II. MOLLUSK SHELLS FROM DEVIL'S MOUTH SITE (41 VV 188)

        Amblema sp.

        Archatinid sp.

        Bulimulus alternatus

        Gastrocopta contracta

        Gastrocopta cristata

        Gastrocopta pellucida hordeacella

        Gastrocopta procera

        Gastrocopta sp.

        Gyraulus parvus

        Hawaiia minuscula

        Helicodiscus parallelus

        Helicodiscus singleyanus

        Helisoma anceps

Devil's Mouth Site (cont'd)

Helisoma _trivolvis_

Helisoma sp.

Lampsilis sp.

Physa _anatina_

Physa _gyrina_

Pisidium _nitidum_

Planorbis sp.

Polygyra _texasiana_

Proptera sp.

Punctum _vitreum_

Pupoides _albilabris_

Succinea sp.

Tropicorbis _obstructus_

Intrasite Distribution, Area A:

 Stratum 1:

   Bulimulus _alternatus_

   Bulimulus _dealbatus_

   Gastrocopta _pellucida_ _hordeacella_

   Hawaiia _minuscula_

   Succinea sp.

 Stratum 2:

   Gastrocopta _pellucida_ _hordeacella_

   Hawaiia _minuscula_

   Succinea sp.

Devil's Mouth Site (cont'd)

Stratum 3:

> Bulimulus alternatus
>
> Gastrocopta cristata
>
> Gastrocopta pellucida hordeacella
>
> Gastrocopta procera
>
> Hawaiia minuscula
>
> Helisoma sp.
>
> Pisidium nitidum
>
> Pupoides albilabris
>
> Succinea sp.

Stratum 4:

> Bulimulus sp.
>
> Gastrocopta cristata
>
> Gastrocopta pellucida hordeacella
>
> Gastrocopta procera
>
> Hawaiia minuscula
>
> Helicodiscus parallelus
>
> Physa gyrina
>
> Planorbis sp.
>
> Polygyra texasiana
>
> Succinea sp.

Devil's Mouth Site (cont'd)

Stratum 5:

Bulimulus sp.

Gastrocopta contracta

Gastrocopta cristata

Gastrocopta pellucida hordeacella

Gastrocopta procera

Hawaiia minuscula

Pupoides albilabris

Succinea sp.

Immature *achatinid* (probably *Subulina* or *Lamellaxis*)

Stratum 6:

Archatinid (?) sp.

Bulimulus sp.

Gastrocopta pellucida hordeacella

Gastrocopta procera

Hawaiia minuscula

Helisoma anceps

Punctum vitreum

Succinea sp.

Devil's Mouth Site (cont'd)

Stratum 7:

Gastrocopta cristata

Gastrocopta procera

Pupoides albilabris

Tropicorbis obstructus

Stratum 8:

Gastrocopta procera

Hawaiia minuscula

Stratum 9:

Bulimulus dealbatus

Bulimulus sp.

Gastrocopta cristata

Gastrocopta pellucida hordeacella

Gastrocopta procera

Gyraulus parvus

Hawaiia minuscula

Polygyra texasiana

Pupoides albilabris

Succinea sp.

Devil's Mouth Site (cont'd)

Stratum 10:

Gastrocopta procera

Gastrocopta sp.

Hawaiia minuscula

Helisoma sp.

Physa anatina

Pupoides albilabris

Stratum 11:

Gastrocopta pellucida hordeacella

Gastrocopta sp.

Gyraulus parvus

Hawaii minuscula

Planorbis (?) sp.

Pupoides albilabris

Succinea sp.

Stratum 12:

Bulimulus sp.

Hawaiia minuscula

Succinea sp.

Devil's Mouth Site (cont'd)

Stratum 13:

Archatinid (?) sp.

Gastrocopta pellucida hordeacella

Helisoma trivolvis

Physa anatina

Pisidium nitidum

Pupoides albilabris

Stratum 14:

Gastrocopta sp.

Gyraulus parvus

Hawaiia minuscula

Pupoides albilabris

Stratum 15:

Gastrocopta procera

Hawaiia minuscula

Helicodiscus singleyanus

Intrasite Distribution Area B (all depths below surface)

Surface to 0.5 ft.:

Amblema sp.

Bulimulus alternatus

Devil's Mouth Site (cont'd)

1.0 to 1.5 ft.:

Amblema (?) sp.

Bulimulus alternatus

1.5 to 2.0 ft.:

Bulimulus alternatus

Proptera sp.

2.5 to 4.0 ft.:

Amblema (?) sp.

Bulimulus alternatus

Lampsilis sp.

4.0 to 4.5 ft.:

Bulimulus alternatus

Lampsilis (?) sp.

4.5 to 5.0 ft.:

Bulimulus alternatus

Lampsilis (?) sp.

6.5 to 7.0 ft.:

Amblema (?) sp.

Bulimulus alternatus

Proptera sp.

7.0 to 7.5 ft.:

Bulimulus alternatus

Lampsilis (?) sp.

Proptera (?) sp.

Devil's Mouth Site (cont'd)

7.5 to 8.0 ft.:

Bulimulus _alternatus_

Proptera sp.

8.0 to 9.0 ft.:

Bulimulus _alternatus_

III.   MOLLUSK SHELLS FROM BONFIRE SHELTER (41 VV 218)

Bulimulus _alternatus_

Bulimulus _schiedeanus_

Bulimulus sp.

Catinella _vermeta_

Discus _cronkhitei_

Hawaiia _minuscula_

Succinea sp.

Intrasite Distribution:

Fiber Layer:

Bulimulus _alternatus_

Bulimulus _schiedeanus_

Bulimulus sp.

Discus _cronkhitei_

Succinea sp.

Bone Bed 2:

Bulimulus _alternatus_

Bonfire Shelter (cont'd)

Zone I:

Bulimulus alternatus

Catinella vermeta

Hawaiia minuscula

IV. MOLLUSK SHELLS FROM DEVILS ROCKSHELTER (41 VV 264)

Bulimulus alternatus

Bulimulus dealbatus

Bulimulus sp.

Catinella vermeta

Durangonella sp.

Gastrocopta pellucida hordeacella

Gastrocopta procera

Gastrocopta tappiniana

Hawaiia minuscula

Heliocodiscus parallelus

Helisoma trivolvis

Laevapex fuscus (Ferrissia rivularis)

Lamellaxis sp.

Physa anatina

Polygyra dorfeuilliana

Polygyra texasiana

Polygyra sp.

Pupoides albilabris

Sphaerium striatinum

Devils Rockshelter (cont'd)

    Succinea avara

    Succinea sp.

Intrasite Distribution:

  Stratum VII:

    Bulimulus alternatus

    Catinella vermeta

    Gastrocopta pellucida hordeacella

    Hawaiia minuscula

    Polygyra sp.

    Succinea avara

  Stratum V:

    Bulimulus dealbatus

    Gastrocopta pellucida hordeacella

    Gastrocopta procera

    Gastrocopta tappaniana

    Hawaiia minuscula

    Laevapex fuscus (Ferrissia rivularia)

    Physa anatina

    Polygyra dorfeuilliana

    Polygyra texasiana

    Pupoides albilabris

    Succinea sp.

Devils Rockshelter (cont'd)

Stratum IV:

Bulimulus dealbatus

Gastrocopta pellucida hordeacello

Gastrocopta procera

Hawaiia minuscula

Helicodiscus parallelus

Physa anatina

Polygyra sp.

Pupoides albilabris

Stratum III:

Bulimulus sp.

Gastrocopta procera

Gastrocopta tappaniana

Hawaiia minuscula

Pupoides albilabris

Succinea sp.

Stratum II:

Durangonella sp.

Gastrocopta pellucida hordeacella

Gastrocopta procera

Hawaiia minuscula

Helisoma trivolvis

242

Devils Rockshelter (cont'd)

Stratum II (cont'd)

Lamellaxis sp.

Polygyra texasiana

Sphaerium striatinum

Succinea sp.

Stratum I:

Bulimulus sp.

Gastrocopta pellucida hordeacella

Hawaiia minuscula

Polygyra texasiana

Succinea sp.

SUMMARY

According to all publications pertaining to the current molluscan fauna in the Amistad area and in adjacent counties, the species collected in the excavations are strikingly similar to the Recent species. However, it must be remembered that our information on Recent gastropods in the Amistad region is far from complete. In order to get a comparative "picture" of the fossil versus Recent invertebrate fauna, a great deal more collecting must be done for both faunas.

In collecting fossil shells, among which are minute forms, it is essential to use the washing technique described by Hibbard (1949). Essentially, this technique consists of placing the matrix in large screen sieves in which the mesh is sufficiently minute to hold back shells of less than 1 mm. in size. By gently washing the matrix and by straining sediments, one can recover the species and determine their relative frequencies of occurrence. This quantitative study is extremely important in the use of shells as paleoecological indicators.

    My chief recommendation for further work is that
extensive and intensive collecting for both Recent and
fossil shells be done by experienced collectors. It is
only after these more adequate collections have been made
and analyzed that we can be in a position to make specific, relevant statements concerning the paleoecology of
the Amistad area.

REFERENCES CITED

Anderson, Roger Y.
    1955. Pollen Analysis, a Research Tool for the Study
        of Cave Deposits. American Antiquity, Vol. 21,
        No. 1: 84-85.

Antevs, Ernst
    1955. Geologic-climatic Dating in the West. American
        Antiquity, Vol. 20, No. 4: 317-335.

    1962. Late Quaternary Climates in Arizona. American
        Antiquity, Vol. 28, No. 2: 193-198.

Baird, S. F., and C. Girard
    1853. Catalogue of North American Reptiles in the
        Museum of the Smithsonian Institution. Pt. 1.
        Serpents. Smithsonian Miscellaneous Collection,
        Vol. 2, Article 5.

Baker, R. H.
    1953. The Pocket Gophers (genus *Thomomys*) of Coahuila,
        Mexico. University of Kansas Publication,
        Museum of Natural History, No. 5: 499-514.

Bent, A. M., and H. E. Wright, Jr.
    1963. Pollen Analysis of Surface Materials and Lake
        Sediments from the Chuska Mountains, New Mexico,
        Bulletin of Geological Society of America,
        Vol. 74, No. 4: 491-500.

Blair, W. Frank
    1950. The Biotic Provinces of Texas. The Texas
        Journal of Science, Vol. 2, No. 1: 93-117.

    1958. Distribution Pattern of Vertebrates in the
        Southern States in Relation to Past and Present
        Environment. Zoogeography, American Association
        for the Advancement of Science: 433-468.

Braun, E. Lucy
    1955. The Phytogeography of Unglaciated Eastern
        United States and Its Interpretation. Botanical
        Review, Vol. 21: 297-375.

246

Brown, B. C.
    1950.   An annotated checklist of the reptiles and
            amphibians of Texas. Baylor University

Bray, W. L.
    1905.   Vegetation of the Sotol Country in Texas.
            University of Texas Bulletin 60.

Bryan, A. L., and Ruth Gruhn
    1964.   Problems Relating to the Neothermal Climatic
            Sequence. American Antiquity, Vol. 29, No. 3:
            307-315.

Butler, C. T.
    1948.   A West Texas Rock Shelter. Unpublished M.A.
            Thesis on file at The University of Texas.

Callen, E. O.
    1960.   A Prehistoric Diet Revealed in Coprolites.
            The New Scientist, August, 7th Issue: 35-40.

Clisby, K. H., and P. B. Sears
    1956.   San Augustin Plains-Pleistocene Climatic
            Changes in New Mexico, U.S.A. Veröff, Geo-
            botanisches Institut Rübel 34: 21-26.

Conant, R.
    1958.   A Field Guide to Reptiles and Amphibians of the
            United States and Canada East of the 100th
            Meridian. Houghton-Mifflin Co.

Davenport, J. Walker
    1938.   Archeological Exploration of Eagle Cave, Langtry,
            Texas. Witte Memorial Museum, Bulletin No. 4.

Davis, Margaret B.
    1963.   On the Theory of Pollen Analysis. American
            Journal of Science, Vol. 261, December: 897-912.

Davis W. B.
    1940.   Distribution and Variation of Pocket Gophers
            (genus Geomys) in the Southwestern United States.
            Bulletin, Texas Agricultural Experiment Station,
            No. 590.

Deevey, E. S.
    1949.   Biogeography of the Pleistocene. Bulletin of the
            Geological Society of America, Vol. 60: 1315-
            1416.

Dibble, David S.
    1965.   Bonfire Shelter: A Stratified Bison Kill Site in
            the Amistad Reservoir Area, Val Verde County,
            Texas. Miscellaneous Papers, No. 5, Texas
            Archeological Salvage Project.

Epstein, Jeremiah F.
    1963.   Centipede and Damp Caves: Excavation in Val Verde
            County, Texas, 1958. Bulletin of the Texas
            Archeological Society, Vol. 33 (for 1962):
            1-129.

Erdtman, G.
    1957.   Pollen and Spore Morphology, Plant Taxonomy.
            Ronald Press Co.

Faegri, Knut, and Johs. Iversen
    1964.   Textbook of Pollen Analysis. Hafner Publishing Co.

Fenneman, Nevin M.
    1931.   Physiography of Western United States. McGraw-
            Hill Book Co., Inc.

Gehlbach, F. R., and J. K. Baker
    1962.   Kingsnake Allied with Lampropeltis mexicana:
            Taxonomy and Natural History. Copeia: 291-300.

Goldman, E. A.
    1936.   New pocket gophers of the genus Thomomys. Journal
            of the Washington Academy of Science, Vol. 26,
            No. 11: 111-120.

Gould, F. W.
    1962.   Texas Plants - A Checklist and Ecological Sum-
            mary. Texas Agricultural Experimental Station,
            The Agricultural and Mechanical College of
            Texas.

Graham, Alan
    1965.   Origin and Evolution of the Biota of Southeastern
            North America: Evidence From the Fossil Plant
            Record. Evolution, Vol. 18, No. 4: 571-585.

Graham, John Allen, and William A. Davis
    1958. Appraisal of the Archeological Resources of
           Diablo Reservoir, Val Verde County, Texas.
           Report prepared by the Archeological Salvage Field
           Office, Austin, Texas. U.S. National Park Service.

Gray, Jane, and Watson Smith
    1962. Fossil Pollen and Archaeology. Archaeology, Vol.
           15, No. 1: 16-26.

Hafsten, Ulf
    1961. Pleistocene Development of Vegetation and Climate
           in the Southern High Plains as Evidenced by
           Pollen Analysis. Paleoecology of the Llano
           Estacado (F. Wendorf ed.), Fort Burgwin Research
           Center, Publication No. 1, Museum of New Meaico
           Press.

Hall, E. R., and K. R. Kelson
    1959. The Mammals of North American. 2 Vols. Roland
           Press.

Hevly, Richard H.
    1964. Pollen Analysis of Quaternary Archaeological and
           Lacustrine Sediments from the Colorado Plateau.
           PhD. Dissertation on file at the University of
           Arizona.

Hevly, R. H., and Peter J. Mehringer, Jr., and H. G. Yocum
    1965. Studies of the Modern Pollen Rain in the Sonoran
           Desert, Journal of The Arizona Academy of Science,
           Vol. 3, No. 3: 123-135.

Hevly, R. H., and P. S. Martin
    1961. The Geochronology of Pluvial Lake Cochise I.
           Pollen Analysis of Shore Deposits. Journal of the
           Arizona Academy of Science, Vol. 2: 24-31.

Hibbard, Claude W.
    1949. Techniques of Collecting Microvertebrate fossils.
           Contributions of the Museum of Paleontology,
           University of Michigan, Vol. 8: 7-19.

Holden, W. C.
    1937. Excavation of Murrah Cave. Bulletin of the Texas
           Archeological and Paleontological Society, Vol.
           9: 48-73.

Hutchinson, G. E., Ruth Patrick, and Edward S. Deevey
    1958.   Sediments of Lake Patzcuaro, Michoacan, Mexico.
            Geological Society of America Bulletin, Vol. 67:
            1491-1505.

Irwin, Henry, and Elso S. Barghoorn
    1965.   Identification of the Pollen of Maize, Teosinte
            and Tripsacum by Phase Contrast Microscopy.
            Botanical Museum Leaflets, Vol. 21, No. 2: 37-57.

Iversen, Johs.
    1941.   Land Occupation in Denmark's Stone Age. Axel
            Sandal Co.

Jackson, A. T.
    1938.   Picture Writing of Texas Indians. Anthropological
            Papers, No. 2, University of Texas Publication,
            No. 3809.

Jennings, Jesse D. (editor)
    1956.   The American Southwest: A Problem in Cultural
            Isolation (In Seminars in Archaeology: 1955,
            Robert Wauchope, ed.). Memoirs of the Society
            for American Archaeology, No. 11.

Jennings, Jesse D., and Edward Norbeck
    1955.   Great Basin Prehistory: A Review. American
            Antiquity, Vol. 21, No. 1: 1-11.

Johnson, LeRoy, Jr.
    1961.   The Devil's Mouth Site: A River Terrace Midden,
            Diablo Reservoir, Texas. Bulletin of the Texas
            Archeological Society, Vol. 30: 253-285.

    1963.   Pollen Analysis of Two Archeological Sites at
            Amistad Reservoir, Texas. The Texas Journal of
            Science, Vol. 15, No. 2: 225-230.

    1964.   The Devil's Mouth Site: A Stratified Campsite
            at Amistad Reservoir, Val Verde County, Texas.
            Archeology Series, No. 6, Department of Anthro-
            pology, The University of Texas.

Johnston, M. C.
    1955.   Vegetation of the Eolian Plain and Associated
            Coastal Features of Southern Texas. PhD. Dis-
            sertation, The University of Texas.

Kapp, Ronald O.
    1965.   Illinoian and Sangamon Vegetation in Southwestern
            Kansas and Adjacent Oklahoma. Contributions from
            the Museum of Paleontology, Vol. XIX, No. 14:
            167-255.

Kelley, J. Charles, T. N. Campbell, and Donald J. Lehmer
    1940.   The Association of Archeological Materials With
            Geological Deposits in the Big Bend Region of
            Texas. Sul Ross State Teachers College Bulletin,
            Vol. 21, No. 3.

Kirkland, Forrest
    1937.   A Study of Indian Pictures in Texas. Bulletin of
            the Texas Archeological and Paleontological
            Society, Vol. 9: 89-119.

    1938.   A Description of Texas Pictographs. Bulletin of
            the Texas Archeological and Paleontological
            Society, Vol. 10: 11-39.

    1939.   Indian Pictographs in the Dry Shelters of Val
            Verde County, Texas. Bulletin of the Texas
            Archeological and Paleontological Society, Vol.
            11: 47-76.

Kremp, Gerhard O. W.
    1965.   Morphological Encyclopedia of Palynology.
            University of Arizona Press.

Laudermilk, J. D., and P. A. Munz
    1934.   Plants in the Dung of Nothrotherium from Gypsum
            Cave, Nevada. Carnegie Insitution of Washington,
            Publication No. 453: 31-37.

Little, Elbert L., Jr.
    1950.   Southwestern Trees: A Guide to the Native Species
            of New Mexico and Arizona. Agricultural Handbook,
            No. 9, U.S. Department of Agriculture.

Martin, George C.
    n.d.   The Big Bend Basket Maker. Witte Memorial
            Museum, No. 1.

    1933.   Archaeological Exploration of Shumla Caves.
            Witte Memorial Museum Bulletin, No. 3.

Martin, Paul S.
    1963.   The Last 10,000 Years. A Fossil Pollen Record
            of the American Southwest. The University of
            Arizona Press.

Martin, Paul S., and B. F. Harrell
    1957.   Pleistocene History of Temperate Biotas in Mexico
            and Eastern United States. Ecology, Vol. 38:
            468-480.

Martin, P.S., B. Sabels, and D. Shutler
    1961.   Rampart Cave Coprolite and Ecology of the Shasta
            Ground Sloth. American Journal of Science, Vol.
            259: 102-127.

Martin, Paul S., and Floyd W. Sharrock
    1964.   Pollen Analysis of Prehistoric Human Feces: A
            New Approach to Ethnobotany. American Antiquity,
            Vol. 30, No. 2: 168-180.

Mc Dougall, W. B., and Omer E. Sperry
    1951.   Plants of Big Bend National Park. U. S. Govern-
            ment Printing Office.

Milstead, W. W., J. S. Mecham, and H. McClintock
    1950.   The amphibians and reptiles of the Stockton
            Plateau in northern Terrell County, Texas. The
            Texas Journal of Science, Vol. 2, No. 4: 543-562.

Milstead, W. W., and D. W. Tinkle
    1959.   Notes on the Porcupine (Eretizon dorsatum) in
            Texas. Southwestern Naturalist, Vol. 3, Nos.
            1-4: 236-237.

Muller, C. H.
    1937.   Vegetation in the Chisos Mountains, Texas. Trans-
            Texas Academy of Science, Vol. 20: 1-31.

Nelson, E. W., and E. A. Goldman
    1934.   Revision of the Pocket Gophers of the Genus Crato-
            geomys. Proceeding of the Biological Society of
            Washingon, Vol. 47: 135-153.

Nunley, John P., Lathel F. Duffield, and Edward B. Jelks
    1965.   Excavations at Amistad Reservoir, 1961 Season.
            Miscellaneous Papers, No. 3. Texas Archeological
            Salvage Project.

Palmer, E. J.
  1928. Leaves From a Collector's Notebook. Journal of
         Arnold Arboretum, Vol. 9, No. 4: 174-178.

Parsons, Mark L.
  1965. 1963 Test Excavations at Fate Bell Shelter,
         Amistad Reservoir, Val Verde County, Texas.
         Miscellaneous Papers, No. 4, Texas Archeological
         Salvage Project.

Patton, Thomas H.
  1963. Fossil Vertebrates from Miller's Cave, Llano
         County, Texas. Bulletin 7, Texas Memorial Museum,
         The University of Texas.

Pearce, J. E., and A. T. Jackson
  1933. A Prehistoric Rock Shelter in Val Verde County,
         Texas. Anthropological Papers, Vol. 1, No. 3,
         University of Texas Publication, No. 3327.

Pearson, F. J., Jr., E. Mott Davis, M. A. Tamers, and Robert
  W. Johnstone
  1965. University of Texas Radiocarbon Dates III. Radio-
         carbon, Vol. 7: 296-314.

Potzger, J. E., and B. C. Tharp
  1947. Pollen Profile From a Texas Bog. Ecology, Vol. 28:
         274-280.

  1954. Pollen Study of Two Bogs in Texas. Ecology, Vol.
         35: 462-466.

Prewitt, Elton R.
  1966. A Preliminary Report on the Devils Rockshelter
         Site, Val Verde County, Texas. The Texas Journal
         of Science, Vol. 18, No. 2 (in press).

Ross, Richard E.
  1965. The Archeology of Eagle Cave. Papers of the Texas
         Archeological Salvage Project, No. 7, Texas
         Archeological Salvage Project.

Sayles, E. B.
  1935. An Archeological Survey of Texas. Medallon
         Papers, No. 17. Gila Pueblo.

Schoenwelter, James
    1960. Pollen Analysis of Sediments From Matty Wash.
           M.A. Thesis on file at the University of Arizona.

Schoenwelter, James, and Frank Eddy
    1964. Alluvial and Palynological Reconstruction of
           Environments, Navajo Reservoir District. Museum
           of New Mexico Papers in Anthropology, No. 13.

Scheutz, Mardith K.
    1956. An Analysis of Val Verde County Cave Material;
           Bulletin of the Texas Archeological Society, Vol.
           27: 129-160.

    1961. An Analysis of Val Verde County Cave Material:
           Part II. Bulletin of the Texas Archeological
           Society, Vol. 31: 167-205.

    1963. An Analysis of Val Verde Cave Material: Part III.
           Bulletin of the Texas Archeological Society, Vol.
           33: 131-165.

Scultz, Ellen D.
    1928. Texas Wild Flowers. Laidlaw Bros. Publishers.

Scudday, J. F.
    1965. *Eleutherodactylus* *latrans* in Terrell County, Texas.
           Southwestern Naturalist, Vol. 10, No. 1: 78.

Sears, Paul B.
    1952. Palynology in Southern North America. I. Archeo-
           logical Horizons in the Basin of Mexico. Geologi-
           cal Society of America Bulletin, Vol. 66: 521-530.

Sears, P. B., and A. Roosma
    1961. A Climatic Sequence From Two Nevada Caves. Ameri-
           can Journal of Science, Vol. 259: 669-678.

Setzler, Frank
    1934. Cave Burials in Southwestern Texas. Explorations
           and Fieldwork of the Smithsonian Institution in
           1933: 35-37.

Slaughter, Bob H.
    n.d.   An Ecological Interpretation of the Brown Sand
           Wedge Local Fauna, Blackwater Draw, New Mexico:
           and a Hypothesis Concerning Late Pleistocene
           Extinction. Paleocology of the Llano Estacado,
           Vol. 2, assembled by Fred Wendorf and J. J. Hester.

254

Slaughter, Bob H., and B. Reed Hoover
    1963. Sulphur River Formation and the Pleistocene Mam-
          mals of the Ben Franklin Local Fauna. Journal of
          the Graduate Research Center, Southern Methodist
          Univeristy, Vol. 31, No. 3: 132-148.

Steeves, N. W., and E. S. Barghoorn
    1959. The Pollen of Ephedra. Journal Arnold Arboretum,
          Vol. 40: 221-255

Suhm, Dee Ann, and Edward B. Jelks (editors)
    1962. Handbook of Texas Archeology: Type Descriptions.
          Special Publication No. 1, Texas Archeological
          Society; Bulletin No. 4, Texas Memorial Museum.

Tamers, M. A., F. J. Pearson, Jr., and E. Mott Davis
    1964. University of Texas Radiocarbon Dates II. Radio-
          carbon, Vol. 6: 138-159.

Tanzer, E. C., E. O. Morrison, and C. Hoffpuir
    1966. New locality records for amphibians and reptiles
          in Texas. Southwestern Naturalist, Vol. 11, No. 1:
          131-132.

Taylor, Herbert C.
    1948. An Archeological Reconnaissance in Northern
          Coahuilla. Bulletin of the Texas Archeological
          and Paleontological Society, Vol. 19: 74-87.

    1949a. A Tentative Cultural Sequence for the Area About
          the Mouth of the Pecos. Bulletin of the Texas
          Archeological Society, Vol. 20: 73-88.

    1949b. The Archeology About the Mouth of the Pecos.
          Unpublished M.A. Thesis on file at the University
          of Texas.

Taylor, W. W.
    1958. Appendix: Archeological Survey of the Mexican
          Part of Diablo Reservoir. (In Appraisal of the
          Archeological Resources of Diablo Reservoir, Val
          Verde County, Texas, by John A. Graham and William
          A. Davis). Mimeographed report prepared by the
          Archeological Salvage Program Field Office, Austin,
          Texas. U. S. National Park Service.

Tharp, B. C.
    1944. The Mesa Region of Texas: an Ecological Study.
    Proceedings of the Texas Academy of Science, Vol.
    27: 81-91.

Thomas, S. J.
    1933. The Archeology of Fate Bell Shelter. Unpublished
    M.A. Thesis on file at The University of Texas.

Traverse, Alfred
    1955. Pollen Analysis of the Brandon Lignite of Vermont.
    United States Department of the Interior.

Tsukada, Matsuo
    1964. Pollen Morphology and Identification. II.
    Cactaceae. Pollen et Spores, Vol. VI, No. 1:
    45-84.

Warnock, B. H.
    1946. The Vegetation of the Glass Mountains, Texas.
    PhD. Dissertation on file at The University of
    Texas.

Wendorf, Fred (editor)
    1961. Paleoecology of the Llano Estacado. Publication
    No. 1, Fort Burgwin Research Center.

Weber, G. L.
    1950. Observations on the Vegetation and Summer Flora of
    the Stockton Plateau in Northeastern Terrell
    County, Texas. The Texas Journal of Science, Vol.
    2, No. 2:

Whitaker, Thomas W.
    1965. Cucurbits and Cultures in the Americas. Economic
    Botany, Vol. 19, No. 4: 344-349.

Whitehead, Donald R.
    1965. Measurement as a Means of Identifying Fossil Maize
    Pollen. Bulletin of the Torrey Botanical Club,
    Vol. 92, No. 1: 7-20.

Wodehouse, R. P.
    1935. Pollen Grains. McGraw-Hill Co.

United States Department of Commerce
    1964. Local Climatological Data for Del Rio, Texas.

.